A GUIDE TO CALCULUS

"Easy & Fun"

Donald T. Willis *a.k.a Skydogg* and Shane A. Willis

A Guide to Calculus

"Easy & Fun"

Published by Spines Publishing Platform

ISBN: 979-8-90418-239-7

Spines

A GUIDE TO CALCULUS

"Easy & Fun"

Donald T. Willis

a.k.a Skydogg

Shane A. Willis

CALCULUS IS MOTION

BONUS: 127 EASY AND FUN EXAMPLE PROBLEMS FOR CALCULUS I, II, AND III

Donald T. Willis *a.k.a Skydogg*
Shane A. Willis – *co-author*

Creek Wischmeyer – *digitizer*
Thomas G. Richards – *editor*
Heidi Hoffman – *digitizer*

TABLE OF CONTENTS

Introduction	5
Problem Solving	12
Critical Thinking	13
Index	29
Definitions	25
Nomenclature	30
Symbols	31
Chapter 1 Prerequisites	49
Chapter 2 Common Mistakes	99
Chapter 3 Limits	109
Chapter 4 Differentiation	137
Chapter 5 Integration	149
Chapter 6 Special Topics	175
Chapter 7 Special Problems	199

Introduction and Index

WHY THIS BOOK?

"A Calculus Guide"

Bonus: 127 easy and fun example problems for calculus 1, 2, & 3

Goal: A primer that quickly refreshes your memory on how to solve problems

(Why): What is the "correct" problem and "steps" to the solution?

Critical Thinking: How do they work? Why do they work?

1.) Define the **correct** problem.

2.) Easy **text** fun

3.) Easy **reference** fun

Introduction

1. Prerequisites
2. Common mistakes
3. Limits 33 Methodologies = Range
4. Differentiation 10 Rules = Rate of Change
5. Integration 7 Algorithms = Quantity
6. Special Topics
7. Special Problems

Please refer to the index for any information on the symbols, definitions, and nomenclatures in the book.

What makes this calculus book so special?

In this book, you will learn how to find the correct problem as step number one. Dare I say the truth of my many years of experience as an American, 80% of the people in our country are working diligently on solving the wrong problem. This book is going to change that.

Step one: Define the **correct** problem.

WHY

W: What is the correct problem?

H: How does the solution work?

Y: Why does the solution work?

Now

Define the correct problem and solve it with accuracy, precision, and simplicity.

My Teaching Philosophy

I believe that it is more instructive to work the same problem several times than to work many problems one time. My last 13 years of teaching calculus have proven to me that my students not only learn but also retain calculus better and longer using this method.

Why Calculus?

Motion.

- Arithmetic, algebra, geometry, and trigonometry all analyze and describe static objects (not in motion).
- Calculus analyzes and describes dynamic objects (in motion).

Calculus equals movement.

Problem-solving methodology

In this book, you will learn:

1. How to find the correct problem.
2. How to clarify the steps to a solution.
3. How to solve the correct problem with accuracy and precision.

If you follow this methodology, the problem will tell you how to solve itself.

Critical Thinking

Critical thinking is the answer to the question "why?"

What is the correct problem?

What are the steps to a solution?

How does it work?

Why does it work?

Donald T. Willis

Math Credentials

Donald earned his engineering degree (BSCE) from West Virginia Tech: license number ET1750. His architectural certification (CDT) from CSI in California. His master's degree (MBA) from LaSalle. His math teacher's license (*OH3078110*) from the Ohio Education Association. He was a bridge design engineer. He did nuclear engineering research for his first eight years in the army (MOS: 01D20). Currently, he is a calculus teacher and author.

Donald was a professional salesman, a district sales manager in the northeast, a regional sales manager in the eastern USA, a branch manager of a manufacturing plant in the northeast, and, as an entrepreneur, was the director of sales in a partnership in California.

He was a Green Beret for (12) years in the western hemisphere (MOS: 18A, B, C, and F). He is a master rated skydiver with 3,135 jumps (USPA license D22449). He is a motivational speaker who has spoken at clients such as Disneyland, The Ohio State University, The Rose Bowl, and many others. As a licensed math teacher in Ohio (license OH3078110), he teaches for outreach programs, developmental disabilities programs, homeschooling, and is authoring a calculus textbook.

Shane Willis

Shane Willis is a musician, composer, and calibration/chemistry technician. He has been performing for over 26 years, specializing in all styles of music, but especially Funk, Fusion Jazz, Metal, and Latin. He started playing professionally at the age of sixteen and afterwards attended the Fort Hayes Career Center and then Capital University, majoring in Jazz Studies. Shane has performed all over the United States and the Caribbean. He has played at jazz festivals such as Litchfield, Ocho Rios, Kansas City Jazz in the Woods, and Clifford Brown, performing acts including Mimi Fox, Othello Molineaux, Carlos Malta, Warren Hill, Julian Vaughn, and Chris Geith. He has also played with the smooth jazz trumpeter, David Wells, and his band "Hook." In 2014, Shane wrote and produced his first CD, "Breakin' Out," featuring trumpet/flugelhorn performed by David Wells. He is currently working on a solo album called "Decades," featuring his original jazz fusion style.

In 2016, Shane started a career in calibrations and chemistry lab work with the company "Nucon International Inc." Within this company, Shane was responsible for calibrating different measurement devices, creating test methods, and using a variety of ASTM protocols, such as iodine removal efficiency and Krypton/Xenon delay in off-gas systems. He recently finished his final certifications to become the head of the cybersecurity division in this company.

Tom G. Richards

Thomas (Tom) Richards, a native of Pittsburgh, PA, was raised in Connellsville, PA and the surrounding old mountains. He joined the Army a week after his 17th birthday and never looked back. Following tours of duty in Korea, Vietnam, and Europe, Tom served the Army and the national intelligence community in Washington, D.C., ending his career in the Army Space Program Office. He retired after 26 years of service as a Chief Warrant Officer (CW04). Tom continued to serve the national intelligence and defense community and mission for another 30 years, finally fully retiring in 2021, after a total of 56 years of service. Along the way, he served his local communities as a Law Enforcement Officer, Firefighter, Emergency Medical Technician and Court Appointed Special Advocate (protecting children in the courts). Tom accumulated a BA in Management from Eckerd College and a BS/MS in Computer Science from The American University. Married to the Author's sister for 54 years and counting, with one son (whom he is very proud of). Tom and his “bride” live between Florida and North Carolina.

Heidi Hoffman

Heidi Hoffman is currently a business student at Sinclair Community College. She graduated in 2023 as valedictorian from Indian Lake High School. She enjoys reading, painting, and spending time with friends.

Creek Wischmeyer

Creek Wischmeyer, 18, is a recent graduate of Indian Lake High School in Lewistown, Ohio. He is currently attending the Lima branch campus of The Ohio State University, majoring in Psychology and minoring in Mathematics and English. Since childhood, Creek has been infatuated with Psychology and the way that minds work. He can empathize to an extraordinary degree, and it was upon this realization that a career in Psychology became a likely possibility for his future. He found his love for math in high school, when he discovered the relaxing and peaceful side of mathematics. This love prompted his decision to help create this textbook.

Creek has been working since age 15 at his local McDonald's. After less than a year, he was enrolled in their Shift Ambassador program and then promoted to Shift Manager upon graduation. His plan is to stay and manage there while completing his degree. While working, Creek overcame a lot of challenges, one being that he was not able to receive a promotion to manager as a minor. This really put a dampener on his tendencies to always strive for more, but he did not let that discourage him and kept pushing through until graduation. This experience is Creek's reason for pursuing big opportunities any time they arise.

Growing up, Creek was very accustomed to being independent, seeing as he is the third oldest of nine children. His natural inclination to assume leadership positions was fostered at a young age because of this. Throughout high school, Creek held titles such as: Spanish Club President, Envirothon Team Captain, Band President, Drumline Captain, Senior Class Secretary, NHS Committee Chair, Student of the Month, and Senior Homecoming Attendant. All of these organizations and activities paved Creek's road to success, and he hopes to continue exploring opportunities like these in the future.

Preface

"I don't get math."

Then, this book is for you.

Yes, it is a book on calculus... But it is also a window into arithmetic, algebra, geometry, and trigonometry. In this book, you will need to memorize 50 formulas. This is because I teach a method of solving problems.

First, define the correct problem.

Second, clarify the steps to the solution.

Third, solve with accuracy and precision. This results in the problem showing you how to solve itself... Making it easy and fun.

Dedication

Key teachers at Linsly Military, Wheeling, West Virginia (Class 61)

Captain Crawford – English

Captain Judy – English

Captain Peters – Rifle Team

Captain Hagwood – Minstrel

Dorothy – Reading Coach

Without these teachers, I would be nothing... Now, I have achieved twelve dreams and am working on number thirteen.

INDEX: SYMBOLS, NOMENCLATURE, AND DEFINITIONS

$0 \pm at\ point$

$a \pm at\ point$

$\infty \pm$

Try ($\pm$ 0.1) Substitute number + or - (0.1) and solve.

Substitute: Substitute the number and solve

Factoring: Factor the expression and substitute a number, and solve.

Conjugate: Multiply the number by the opposite function.

Rationalize: multiply the expression by the radical ÷ radical.

LCD: Find the least common denominator.

Piecewise (4): find the value and the number of the expression.

Union: write an expression below – and above +

Sine (3) $\frac{\sin x}{x}, \frac{x}{\sin x}, \frac{\sin 5x}{5x} = 1$

Cosine (2) $\cos x, \cos x = 1$

Conjugate: multiply the expression by its opposites

$\frac{2+\sqrt{x}}{den.} \cdot \frac{2-\sqrt{x}}{2-\sqrt{x}}$

Rationalize: multiply the expression by $\frac{\sqrt{x}}{\sqrt{x}}$ to get the radical out of the denominator.

Smaller exponent: $\frac{smaller^{x}}{bigger^{x}} = 0$

Same exponent: $\frac{same^{x}}{same^{x}} =$ 1st coefficient

Bigger exponent: $\frac{bigger^{x}}{smaller^{x}} = \infty (or\ DNE)$

Removable · point on the curve

Nonremovable **o** point at the gap

Continuous · point on the curve

Sine: $\lim_{x \to 0} \frac{\sin 2x}{7x} = 1 \cdot \frac{1}{5} = \frac{1}{5}$

Cosine: $\lim_{x \to 0} \frac{\cos 3x - \cos 5x}{x^2} = \frac{5^2 - 3^2}{2} = 8$

L'Hopital's Rule: $\frac{d/dx}{d/dx}$ then substitute and solve.

$f(a) = \lim x \to (a)$ continuous

(**Note**: if $\neq$ then noncontinuous).

$f(x)\ y = f(x)$: function

$f'(x)$: differential

$f^{-1}(x)$: inverse (reverse $x + y$)

$g'(x)$: g function

x' : d/dx

μ' : $d/d\mu$

v' : d/dv

$\lim x \to (a)$

I : integrand

$l\frac{b}{a}$: limit

d/dx : differentiate x

d/dt : differentiate t (time)

$\int_a^b$: integral

$\mu u - \int$: parts formula

$uv - \int v\, du$ Parts Formula

Parts Solution Grid:

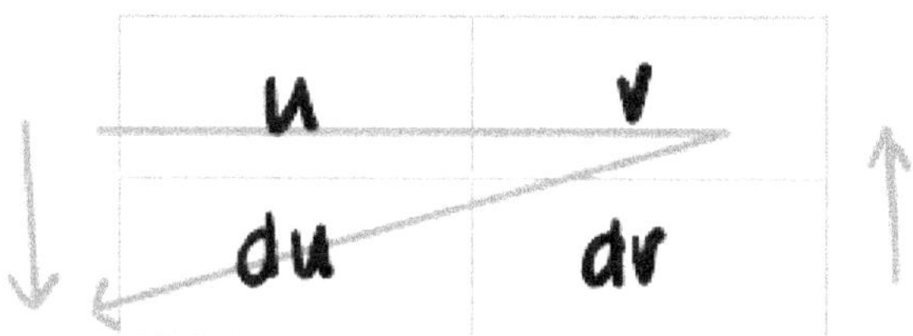

- SISF Trig-Substitution [substitute, SOH CAH TOA, identities, function]
- CRF (PD) } *Complex Rational Function, Polynomial Division*
- LQ SG μ/y' } *Linear Quadratic Subgroup, M-Sub Differential*

---

y function

y' first differential

y'' second differential

Intercepts function (x at $y = 0$)

Extrema; 1st differential (max / min)

- Concavity: 2nd differential POI (inflections)
- Rational integratable: $\frac{x^2}{x^3}$ or $\frac{x^3}{x^4}$
- Irrational: Use polynomial division to change the function to rational ($\frac{x^3}{x^3}$ or $\frac{x^2}{x^2}$)

Definitions in the Index

i. Conjugate: Multiply the expressions by the opposite of the numerator. i.e.:

$$\frac{2+\sqrt{x}}{den.} \cdot \frac{2-\sqrt{x}}{2-\sqrt{x}}$$

ii. Rationalize: Multiply the expression by $\frac{\sqrt{x}}{\sqrt{x}}$ to get the radical out of the denominator.

iii. LCD: Multiply all the denominators by themselves.

i.e.: $\frac{1}{2}+\frac{1}{3}+\frac{1}{4}=\frac{12+8+6}{24}=\frac{26}{24}=\frac{13}{12}=1\frac{1}{12}$

iv. Union (∪): Represents a break in the curve,

i.e.: $[-\infty,-3)\cup[+4,+\infty]$

v. Piecewise (4)

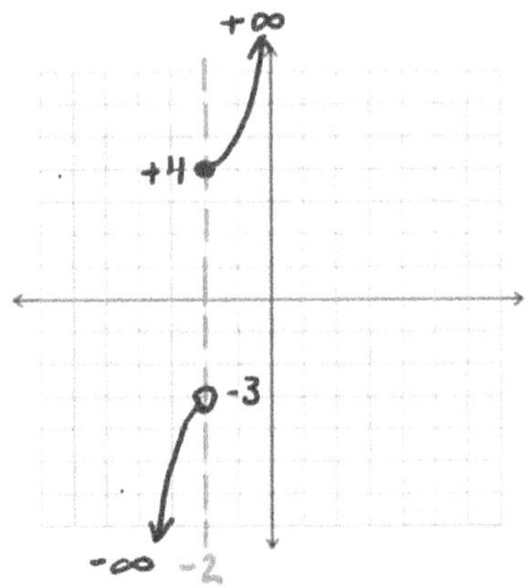

$$\lim_{n \to (-2)^-} = -3$$

$$\lim_{n \to (-2)^+} = +4$$

$$\lim_{n \to (-2)} = DNE$$

$$f(-2) \neq \lim_{n \to (-2)}$$

Not continuous

Note:

To be continuous $f(a) = \lim_{x \to (a)}$ so $\lim_{x \to (a)^-}$ must equal $\lim_{x \to (a)^+}$

Theorem of Calculus

$$function \rightarrow d/dx \rightarrow \int_a^b \rightarrow function$$

$$x^2 \rightarrow 2x \rightarrow \frac{\cancel{2}x^2}{\cancel{2}} \rightarrow x^2$$

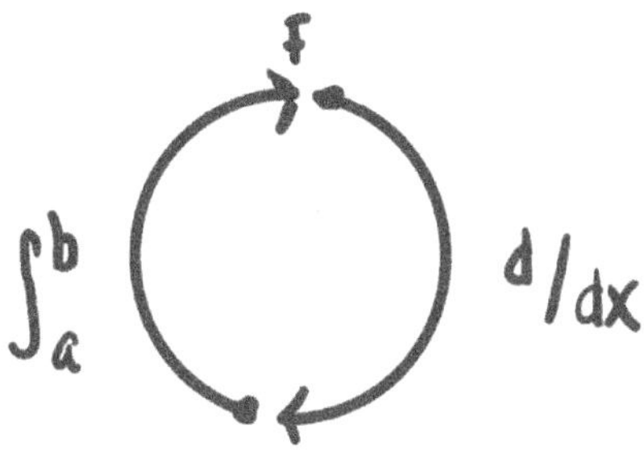

---

i. $f^{-1}(x)$ (Inverse Differentiation): Reverse x and y in the expression, then find the first differential.

ii. SISF (Trigonometry-Substitution): Substitute, identities, SOH CAH TOA, and functions.

iii. CRF (PD) (Partial-Fraction): Complex, rational, functions, and polynomial division.

iv. LQSG μ/y' (Partial-Fractions): Linear, quadratic, substitute, group, μ substitution, and first differential.

v. Concavity: Function (intercepts), 1st differential (max, min, values), 2nd differential (point of inflection POI).

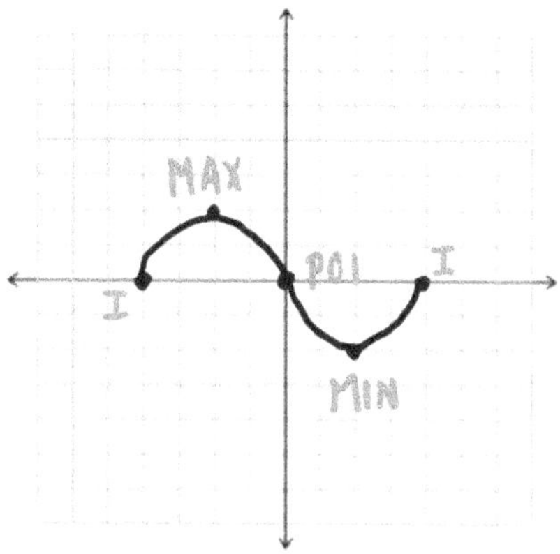

vi. <u>Rational ($\frac{x^2}{x^3}$)</u>: The exponent in the denominator is larger than the exponent in the numerator.

vii. <u>Irrational ($\frac{x^2}{x^2}$or$\frac{x^3}{x^2}$)</u>: The exponent in the numerator is larger than or equal to the exponent in the denominator.

viii. <u>Limit (zoom-in)</u>: Analyze a part of a curve from $x = 1$ and $x = 7$ or $y = 1$ and $y = 7$.

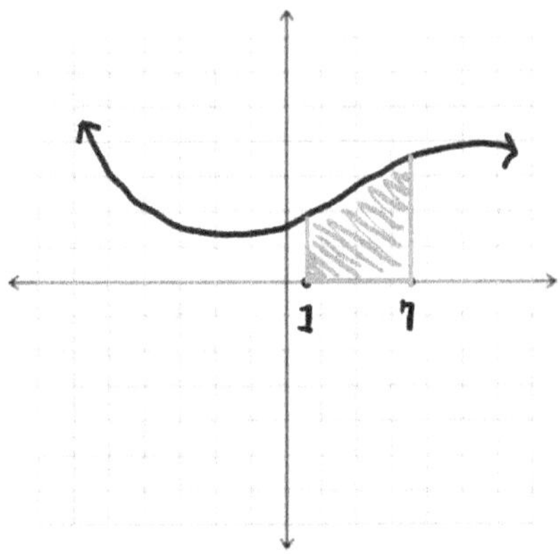

ix. Differential (rate of change): Rate of change of slope, area, or volume over a specific limit (range) of x or y values (re: graph above).

x. Integral (quantity): Finding the quantity (area or volume) under a section of a curve (re: graph above) by dividing it into many small parts, then adding the parts together.

Conversion Tables

Geometry Table

All Triangles

$$\text{Area} = \frac{1}{2}\text{ base} \cdot \text{height}$$

Equilateral Triangle

$$\text{Area} = \frac{\text{side}^2\sqrt{3}}{4}$$

Right Triangle

Pythagorean Theorem:

$$a^2 + b^2 = c^2$$

Parallelogram

$$\text{Area} = \text{base} \cdot \text{height}$$

Trapezoid

$$\text{Area} = \frac{\text{Base}_1 + \text{Base}_2}{2} \cdot \text{height}$$

Circle

$$\text{Area} = \pi r^2$$

$$\text{Circumference} = 2\pi r = \pi d$$

Sector of Circle

Area $= \pi r^2 \left(\frac{\theta}{360°}\right)$

(θ is the central angle)

Arc Length $= 2\pi r \left(\frac{\theta}{360°}\right)$

(Arc Length is the length of the crust)

Sphere

Volume $= \frac{4}{3}\pi r^3$

Surface Area $= 4\pi r^2$

Cone or Pyramid

$$\text{Volume} = \frac{1}{3}A \cdot h$$

(A is the area of the base).

Trig Table

SohCahToa

$$sin\theta = \frac{O}{H} \quad csc\theta = \frac{H}{O}$$

$$cos\theta = \frac{A}{H} \quad sec\theta = \frac{H}{A}$$

$$tan\theta = \frac{O}{A} \quad cot\theta = \frac{A}{O}$$

Degrees and Radians

$2\pi\ radians = 360°$

$\pi\ radians = 180°$

$\frac{\pi}{2}\ radians = 90°$

$\frac{\pi}{3}\ radians = 60°$

$\frac{\pi}{4}\ radians = 45°$

$\frac{\pi}{6}\ radians = 30°$

To convert from radians to degrees, multiply by $\frac{180°}{\pi}$

To convert from degrees to radians, multiply by $\frac{\pi}{180°}$

Identities

Reciprocal identities:

$$csc\theta = \frac{1}{sin\theta}$$

$$sec\theta = \frac{1}{cos\theta}$$

$$cot\theta = \frac{1}{tan\theta}$$

Quotient Identities:

$$tan\theta = \frac{sin\theta}{cos\theta}$$

$$cot\theta = \frac{cos\theta}{sin\theta}$$

Pythagorean Identities:

$$sin^2\theta + cos^2\theta = 1$$

$$tan^2\theta + 1 = sec^2\theta$$

$$1 + cot^2\theta = csc^2\theta$$

Formulas

Half-Angle Formulas:

$$sin^2\theta = \frac{1}{2}(1 - cos2\theta)$$

$$cos^2\theta = \frac{1}{2}(1 + cos2\theta)$$

Double Angle Formulas:

$$sin2\theta = 2\ sin\theta cos\theta$$

$$cos2\theta = 2cos^2\theta - 1$$

Reduction Formulas:

$$\sin(-\theta) = -sin\theta$$

$$\cos(-\theta) = cos\theta$$

$$\tan(-\theta) = -tan\theta$$

Derivative Table

1. $\frac{d}{dx} c = 0$
2. $\frac{d}{dx} x = 1$
3. $\frac{d}{dx} cx = c$
4. $\frac{d}{dx} x^n = nx^{n-1}$
5. $\frac{d}{dx} e^x = e^x$
6. $\frac{d}{dx} \ln x = \frac{1}{x}$
7. $\frac{d}{dx} a^x = a^x \ln a$
8. $\frac{d}{dx} \log_a x = \frac{1}{x} \cdot \frac{1}{\ln a}$
9. $\frac{d}{dx} \sin x = \cos x$
10. $\frac{d}{dx} \cos x = -\sin x$
11. $\frac{d}{dx} \tan x = sec^2 x$
12. $\frac{d}{dx} \cot x = -csc^2 x$
13. $\frac{d}{dx} \sec x = \sec x \tan x$
14. $\frac{d}{dx} \csc x = -\csc x \cot x$
15. $\frac{d}{dx} \arcsin x = \frac{1}{\sqrt{1-x^2}}$
16. $\frac{d}{dx} \arccos x = \frac{-1}{\sqrt{1-x^2}}$
17. $\frac{d}{dx} \arctan x = \frac{1}{1+x^2}$
18. $\frac{d}{dx} \operatorname{arccot} x = \frac{-1}{1+x^2}$
19. $\frac{d}{dx} \operatorname{arcsec} x = \frac{1}{|x|\sqrt{1-x^2}}$
20. $\frac{d}{dx} \operatorname{arccsc} x = \frac{-1}{|x|\sqrt{1-x^2}}$

Integral Table

1. $\int dx = x + C$

2. $\int x^n \, dx = \frac{x^{n+1}}{n+1} + c (n \neq 1)$

3. $\int e^x \, dx = e^x + c$

4. $\int \frac{dx}{x} = ln|x| + c$

5. $\int a^x \, dx = \frac{1}{ln\, a} a^x + c$

6. $\int ln\, x \, dx = x(ln\, x - 1) + C$

7. $\int sin\, x \, dx = -\, cos\, x + c$

8. $\int cos\, x \, dx = sin\, x + C$

9. $\int tan\, x \, dx = -\, ln|cos\, x| + c$

10. $\int cot\, x \, dx = ln|sin\, x| + c$

11. $\int sec\, x \, dx = ln|sec\, x + tan\, x| + c$

12. $\int csc\, x \, dx = -\, ln|csc\, x + cot\, x| + c$

13. $\int sec^2 x \, dx = tan\, x + c$

14. $\int csc^2 x \, dx = -\, cot\, x + C$

15. $\int sec\, x\, tan\, x \, dx = sec\, x + c$

16. $\int \csc x \cot x \, dx = -\csc x + C$

17. $\int \frac{dx}{\sqrt{a^2-x^2}} = \arcsin\frac{x}{a} + c$

18. $\int \frac{dx}{a^2+x^2} = \frac{1}{a}\arctan\frac{x}{a} + c$

19. $\int \frac{dx}{x\sqrt{x^2-a^2}} = \frac{1}{a}\operatorname{arcsec}\frac{|x|}{a} + c$

20. $\int \frac{dx}{x^2-a^2} = \frac{1}{2a}\ln\left|\frac{x-a}{x+a}\right| + c$

Chapter One

Prerequisites

Algebra (9)

F Factor

C Conjugate

R Radicalize

L Least Common Denominator

L Linear

Q Quadratic

P Polynomial

PSD Polynomial Synthetic Division

P Piecewise

U Union

Geometry (7)

A Area

V Volume

C Circle

T Triangle

P Polygon

C Congruent

S Similar

Trigonometry (3)

F Functions

I Identities

U Unit circle

Prerequisites

Arithmetic (See Chapter Two)

Algebra

Geometry

Trigonometry

Chapter One: Prerequisites

This chapter is on the prerequisites of arithmetic, algebra, geometry, and trigonometry. My goal is not to teach you these four subjects, but to review what parts of them you will need to know to successfully do calculus. I am going to do this by showing you the solution to a variety of problems, which you can use as examples to follow.

Algebra

Linear

Slope-intercept Formula: $y = mx + b$

$y = \frac{1}{2}x + 2$

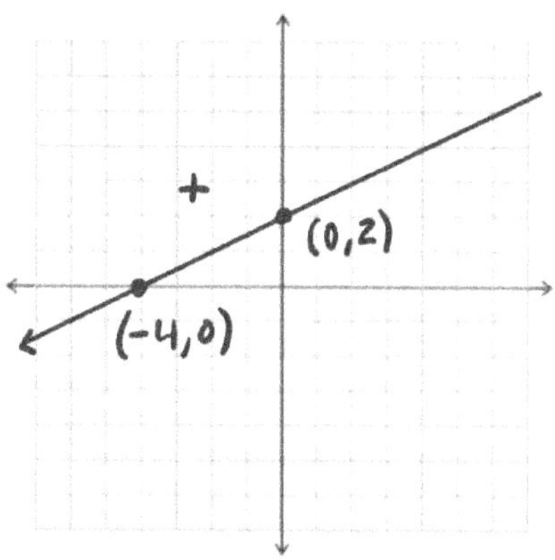

$$m = slope = \frac{y_2 - y_1}{x_2 - x_1} = \frac{2 - 0}{0 - (-4)} = \frac{2}{4} = \frac{1}{2}$$

$$b = y\ intercept\ at\ (x = 0) = 2$$

$$y = mx + 2 = \frac{1}{2} + 2$$

$ax^2 + bx + c = 0$ Quadratic Equation

$x^2 + 5x + 6 = 0$

$(x + 2)(x + 3)$ Factoring

$\boldsymbol{x = -2, -3}$ Answer

$\dfrac{-b \pm \sqrt{b^2 - 4ac}}{2a}$ Quadratic Formula

$$a = 1 \quad b = 5 \quad c = 6$$

$$\frac{-5 \pm \sqrt{5^2 - 4(1)(6)}}{2(1)}$$

$$-5 \pm \sqrt{25 - 24} = \frac{-5 \pm 1}{2}$$

$$-\frac{4}{2}, \frac{-6}{2} = \boldsymbol{x = -2, -3}$$

Difference of Squares

$$x^2 - y^2 = (x - y)(x + y)$$

Difference of Cubes

$$x^3 - y^3 = (x - y)(x^2 + xy + y^2)$$

Sum of Cubes

$$x^3 + y^3 = (x + y)(x^2 - xy + y^2)$$

Polynomial Division

1.) $\frac{64x^3-27}{4x-3}$

Irrational

$$\begin{array}{r|l} & 16x^2 + 12x + 9 \\ \hline 4x-3 & 64x^3 + 0x^2 + 0x - 27 \\ & \underline{64x^3 - 48x^2} \\ & \qquad 48x^2 + 0x \\ & \qquad \underline{48x^2 - 36x} \\ & \qquad\qquad 36x - 27 \\ & \qquad\qquad \underline{36x - 27} \end{array}$$

2.) $\frac{3x^3+2x-11}{x-3}$

Irrational

$$\begin{array}{r|l} & 3x^2 + 9x + 29 + \frac{76}{x-3} \\ \hline x-3 & 3x^3 + 0x^2 + 2x - 11 \\ & \underline{3x^3 - 9x^2} \\ & \qquad 9x^2 + 2x \\ & \qquad \underline{9x^2 - 27x} \\ & \qquad\qquad 29x - 11 \\ & \qquad\qquad \underline{29x - 87} \\ & \qquad\qquad\quad 76 \textit{ remainder} \end{array}$$

Note: Change irrational to rational

Geometry

$Area = length \times width.$

$Volume = length \times width \times height.$

Polygon: 3 or more sides.

Congruent: Same shape, not size.

Two Column Proof

Rules to evaluate two-column proofs:

Given. Side side side. Side angle side. Angle side angle. Side Side Angle (ambiguous). Hypotenuse leg. Congruent parts of congruent triangles are congruent. Reflexive. Transitive.

Trigonometry

$\dot{F}$ $\dot{F}$ S DR DRP LS LC $\dot{UC}$ $\dot{G}$ $\dot{V}$

Function

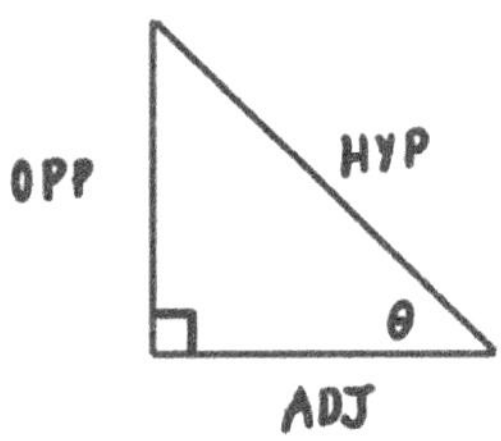

$$sin = \frac{OPP}{HYP}$$

$$cos = \frac{ADJ}{HYP}$$

$$tan = \frac{OPP}{ADJ}$$

Reciprocal

$$sin = \frac{1}{csc} \qquad cos = \frac{1}{sec} \qquad tan = \frac{1}{cot} = \frac{sin}{cos}$$

Pythagorean

$$sin^2 + cos^2 = 1 \quad tan^2 + 1 = sec^2 \quad cot^2 + 1 = csc^2$$

Law of Sines and Cosines

$$\frac{A}{sinA} = \frac{B}{sinB} = \frac{C}{sinC} \qquad a^2 = b^2 + c^2 - 2bccos$$

Graphing

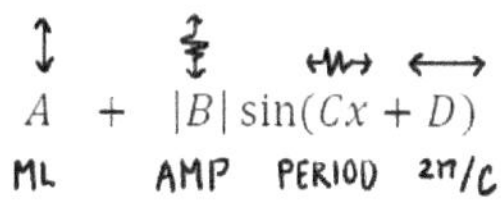

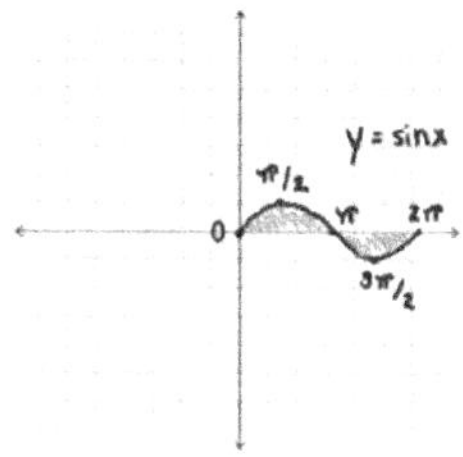

Pause Shift $(Cx + D) = 0$

Trigonometry

Functions

SOH CAH TOA

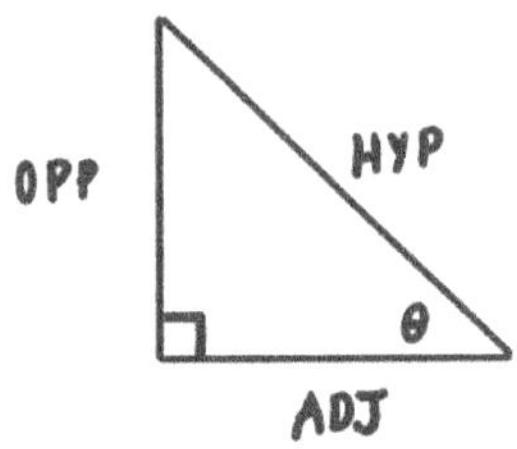

$$\sin\theta = \frac{OPP}{HYP}$$

$$\cos\theta = \frac{ADJ}{HYP}$$

$$\tan\theta = \frac{OPP}{ADJ}$$

Identities

$$\sin = \frac{1}{csc}$$

$$\cos = \frac{1}{sec}$$

$$\tan = \frac{1}{cot} = \frac{sin}{cos}$$

$$\sin^2 \cos^2 = 1$$

$$\tan^2 + 1 = \sec^2$$

$$\cot^2 + 1 = \csc^2$$

Unit Circle

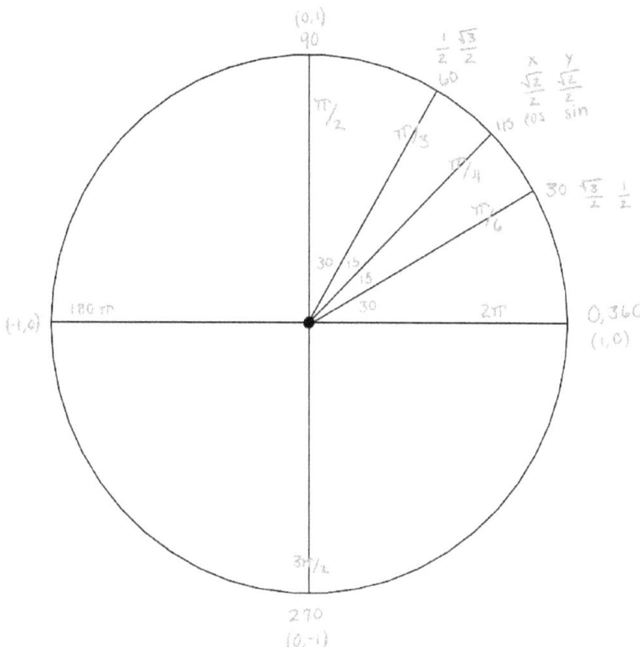

ASTC	
+	+ S
−	+ C
−	+ T
−	−
−	+
+	−

Domain Range

$\sin^{-1} \tan^{-1}(^{\pi}/_{2} - {}^{\pi}/_{2})$

$\cos^{-1}(0, \pi\ \ )$

Degree Radian Polar

$\infty \pm \left[\begin{array}{l} \sin \cos \ (2\pi k) \\ \quad \tan \ (\pi k) \end{array} \right.$ Rad=57.3°

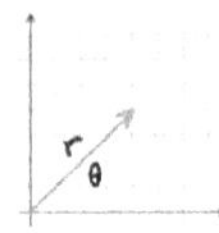

(r, θ)

Graphing

$$A + |B| \sin(Cx + D)$$

ML AMP PERIOD $2\pi/C$

Pause Shift $(Cx + D) = 0$

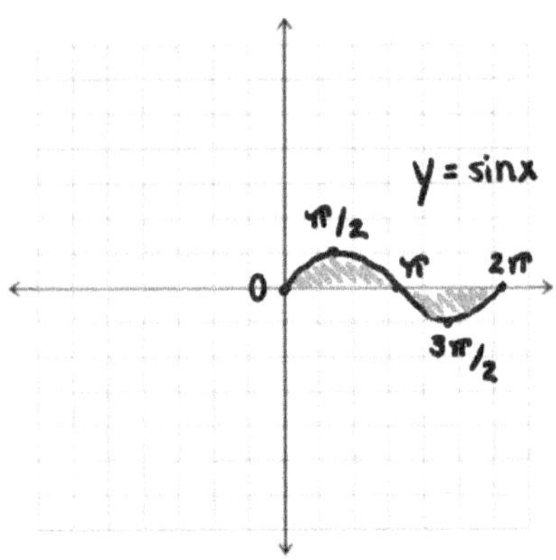

Vectors

1. Law sines: $\frac{A}{sinA} = \frac{B}{sinB} = \frac{C}{sinC}$

2. Law cosines: $a^2 = b^2 + c^2 - 2bccos$

Airplane

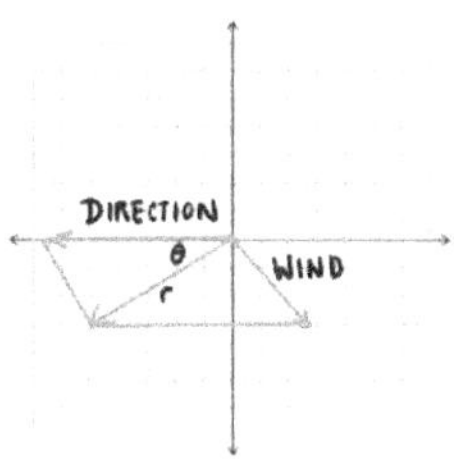

Force

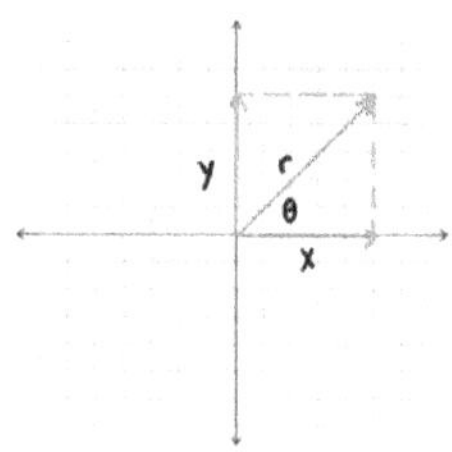

Identities

Reciprocal

$$sin = \frac{1}{csc} \qquad cos = \frac{1}{sec} \qquad tan = \frac{1}{csc} = \frac{sin}{cos}$$

Pythagorean

$$sin^2 + cos^2 = 1 \quad tan^2 + 1 = sec^2 \quad cot^2 + 1 = csc^2$$

Double

$$sin2\theta = 2sin\theta cos\theta \quad cos2\theta = cos^2 - sin^2\theta \quad tan2\theta = \frac{2tan\theta}{1 - tan^2\theta}$$

$$= 2cos^2\theta - 1$$

$$= 1 - 2sin^2\theta$$

$$\tan 2\theta = \frac{2\tan\theta}{1 - \tan^2\theta}$$

Half

$$sin\frac{1}{2}\theta = \pm\sqrt{\frac{1-cos\theta}{2}} \qquad cos\frac{1}{2}\theta = \pm\sqrt{\frac{1+cos\theta}{2}}$$

$$tan\frac{1}{2}\theta = \pm\sqrt{\frac{1 - cos\theta}{1 + cos\theta}} = \frac{1 - cos\theta}{sin\theta} = \frac{sin\theta}{1 + cos\theta}$$

Reducing

$$sin\theta = \sqrt{\frac{1 - cos2\theta}{2}} \quad cos\theta = \sqrt{\frac{1 + cos2\theta}{2}}$$

$$tan\theta = \sqrt{\frac{1 - cos2\theta}{1 + cos2\theta}}$$

Sum difference

$$\sin(x + y) = sinxcosy + sinycosx$$

$$\cos(x + y) = cosxcosy - sinxsiny$$

$$\tan(x + y) = \frac{tanx + tany}{1 - tanxtany}$$

Cofunction

$$sin\theta = \cos(90 - \theta)$$

Unit Circle

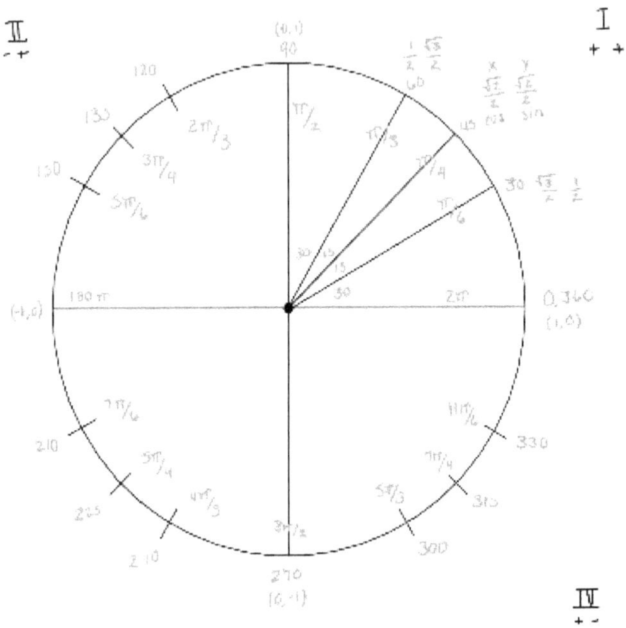
I
+ +
II
- +
III
- -
IV
+ -
(0,1)
90
120
135
150
(-1,0)
180 π
0, 360
(1,0)
2π
210
225
270
(0,-1)
300
315
330
π/2
2π/3
3π/4
5π/6
π/3
π/4
π/6
7π/6
5π/4
4π/3
5π/3
7π/4
11π/6
60
45
30

ASTC	
+	+ S
−	+ C
−	+ T
−	−
−	+
+	−

Domain/Range

$sin^{-1}\left[\frac{\pi}{2}, -\frac{\pi}{2}\right]$

$tan^{-1}\left[\frac{\pi}{2}, -\frac{\pi}{2}\right]$

$cos^{-1}\ [0, \pi]$

$$\infty \pm \begin{bmatrix} \begin{matrix}\text{sin}\\ \text{cos}\end{matrix} & [2\pi k] \\ \text{tan} & [\pi k] \end{bmatrix}$$

Deg / Rad / Polar

$\frac{\theta\,\pi}{180}$ = Rads

Rad = 57.3°

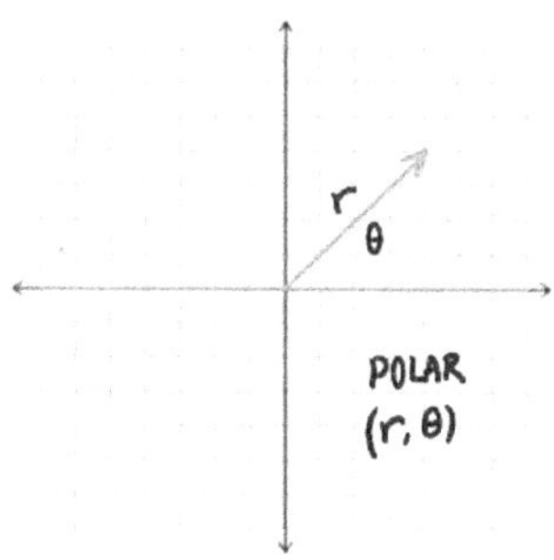

Extra Methodologies

Algebra

1. GCF 2. Polynomials	}	4t Terms (Constant factor tree TI84 Syndiv quadfform)
3. Factoring 4. Guess 5. Universe	}	3 Terms
6. Diff. Sqrs. 7. Sum Diff. Cubes	}	2 Terms
8. Complete Sqr.	}	A = 1
9. Quadratic		
10. Factor Tree		

Sequence	GCF
Terms	CDLCD
1st Coefficient	XX
All Coefficient	Reduce

Equations – 15

$Ax + By + C$

$y = mx + b$

$ax^2 + bx + c$

$a(x - h)^2 + k$

$a(x - p)(x - q)$

$a|x - h| + k$

$x^2 - 6x + 5 \quad 28x^2 - 180x - 112$

$x^2 - y^2 = (x - y)(x + y)$

$x^3 + y^3 = (x - y)(x^2 - xy + y^2)$

$x^3 - y^3 = (x - y)(x^2 + xy + y^2)$

$x = \frac{-b \pm \sqrt{b^2 - 4ac}}{2a}$

$x_v = \frac{-b}{2a}$

$y_v = ax^2 + bx + c$

$b^2 + 4ac$ Discriminant

> 0	2 Roots
$= 0$	1 Root
< 0	No Roots

Factoring

Factoring

$x^2 - 6x + 5$

$(x - 5)(x - 1) = 0$

$x = 5, 1$

Complete Square

$x^2 - 6x + 5$

$x^2 - 6x + 9 = -5 + 9$

$\left(-\frac{6}{2}\right)^2 \rightarrow +9$

$x^2 - 6x + 9 = 4$

$(x - 3)^2 = \pm 2$

$x - 3 = \pm 2$

$x = 5, 1$

Universal

$28x^2 - 180x - 112$

$GCF = 4$

$7x^2 - 45x - 28$

$(7x + \frac{4}{7})(x - \frac{49}{7})4$

$(7x + 4)(x - 7)4$

$7x = 4 \qquad x = 7$

$x = {}^{-4}/_{7}, 7$

$+,-$ (*RSCS*) Absolute, Inequalities $X \div by\ R$

$2|1-x|+1 \geq 3$ Plus   Minus

$\underline{-1} \quad \underline{-1}$ $1-x \geq 1$ $1-x \leq -1$

$\underline{-1} \quad \underline{-1} \qquad \underline{-1} \quad \underline{-1}$

$\frac{2|1-x|}{2} \geq \frac{2}{2}$

$|1-x| \geq 1$ $\frac{-x}{-1} \geq \frac{0}{-1}$ $\frac{-x}{-1} \leq \frac{-2}{-1}$

$x \leq 0$ $x \geq 2$

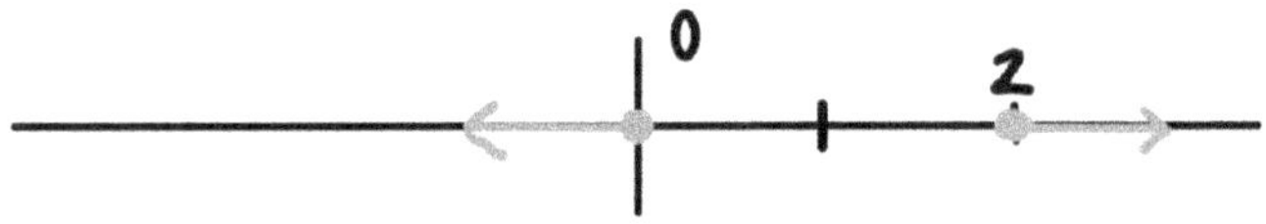

$|x| = 5$ Plus $x = 5$ Minus $x = 5$

$x \geq 5$

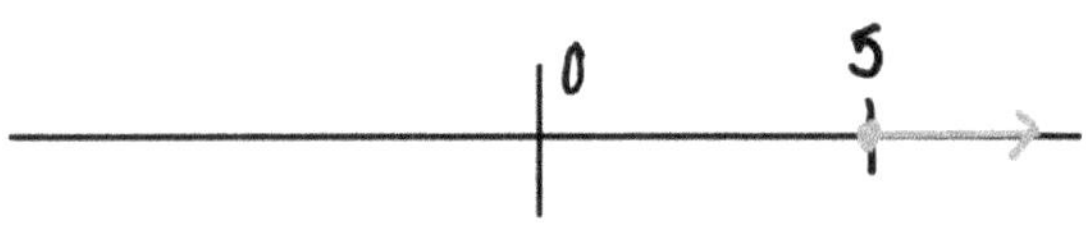

$|x| \geq 5$ Plus $x \geq 5$ Minus $x \leq 5$

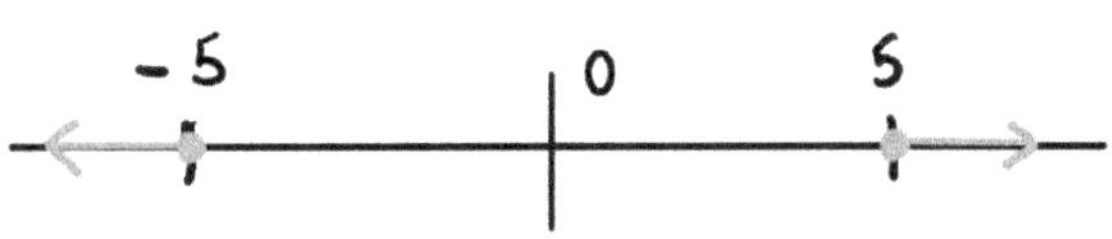

RS: Reverse Symbol   CS: Change Sign

Synthetic Division

Irrational

$$\frac{3x^3+2x-11}{x-3}$$

FT

3

/ \

1 3

$$\frac{3x^3+0x^2+2x-11}{x-3}$$

```
3⌋ | 3 + 0 + 2 − 11
   |_____9__27__87
     3 + 9 + 29 + 76 Remainder
```

$$3x^2+9x+29+\frac{76}{x-3} \therefore \textit{Not a Root}$$

```
1⌋ | 3 + 0 + 2 − 11
   |_____3__3___3
     3   3   5  −6
```

$$3x^2+3x+5-\frac{6}{x-3} \therefore \textit{Not a Root}$$

Simplifying Radicals

$\sqrt[4]{x^{15}} = x^3\sqrt[4]{x^3}$ $\quad\quad (x^{\frac{15}{4}} = x^3\sqrt[4]{x^3})$

$\uparrow$

$15 - 12 = 3$

$$\sqrt[2]{50x^3y^5} = \sqrt{25}\sqrt{2}\sqrt{x^2}\sqrt{x}\sqrt{y^4}\sqrt{y} = 5xy^2\sqrt{2xy}$$

$$\sqrt[2]{72x^4y^5z^6} = \sqrt{36}\sqrt{2}\sqrt{x^4}\sqrt{y^4}\sqrt{y}\sqrt{26}$$
$$= 6x^2\ y^2\ z^3\ \sqrt{2y}$$

$\uparrow$ odd

$$\sqrt[2]{3^3} = 3^{\frac{3}{2}} = \sqrt{3\cdot 3\cdot 3} = \sqrt{9\cdot 3} = 3\sqrt{3}$$

Radicals

$\sqrt{504} = 22.45$

2	252
2	126
2	63
3	21
3	7

$\sqrt{2 \cdot 2 \cdot 2 \cdot 3 \cdot 3 \cdot 7}$

$\sqrt{4} \cdot \sqrt{2} \cdot \sqrt{9} \cdot \sqrt{7}$

$2 \cdot 3 \cdot \sqrt{2} \cdot \sqrt{7}$

$6\sqrt{14} = 22.45$

Remember:

$\sqrt{2} = 1.414$

$\sqrt{3} = 1.732$

$\sqrt{5} = 2.236$

Exponents

$$\frac{(3^2)^3 \cdot 3^6}{3^4} = \frac{3^6 \cdot 3^6}{3^4} = \frac{3^{12}}{3^4} = 3^{12-4} = 3^8$$

Scientific Notation

$$\frac{4x10^{10}}{2x10^5} = 2x10^{10-5} = 2x10^5$$

OOO PEMDAS

(2 Parts: Number, Power) Methodology

OOO: Order of Operation Step / Left, Step / Left

et.

$6 \div 2(1+2)$ Wrong

$6 \div 2 \times 3$ $6 \div 6 = 1$

3×3

9

Exponents

$x^0 = 1$

$4^{-2} = \frac{1}{4^2} = \frac{1}{16}$

$x^{2/3} = (\sqrt[3]{x})^2 = \sqrt[3]{x^2}$

$x^2 \cdot x^3 = x^5$

$\frac{x^5}{x^3} = x^2$

$(x^2)^3 = x^6$

$\sqrt{0} = 0$ $\sqrt{1} = 1$ $\sqrt{-1} = i$

$\sqrt[3]{\sqrt[4]{x}} = \sqrt[12]{x}$

$\sqrt[3]{x^3} = x^{2/3} = x$

$\frac{2}{\sqrt{3}} \cdot \frac{\sqrt{3}}{\sqrt{3}} = \frac{2\sqrt{3}}{3}$ <u>Rationalize</u>

$\frac{2}{\sqrt{3}-1} \cdot \frac{\sqrt{3}+1}{\sqrt{3}+1}$ <u>Conjugate</u>

$\frac{2(\sqrt{3}+1)}{3-1} = \frac{2(\sqrt{3}+1)}{2}$

$\sqrt{3} + 1$

Series

$\sum_{u=1}^{6} 4u = 84$ Calculus

$4(1 + 2 + 3 + 4 + 5 + 6)$ APAB

$4(21)$ APBC

84

- Convergence $(\frac{1}{2} + \frac{1}{4} + \frac{1}{8} + \frac{1}{16} \ldots point.)$

  Approach a point.

- Divergence $(2 + 4 + 8 + 16 \ldots \infty.)$

  Goes to infinity.

T-Substitution. P-Fraction.

y = Function (I) y' = Extrema (mm)

y" = Concavity (POI)

Theorem of Calculus

$$(function \rightarrow {}^{d}/_{dx} \rightarrow \int_a^b \rightarrow function)$$

Infinity

$\lim_{x \to \infty \pm}$ **(SSB)**

<u>Smaller</u>

$\frac{x^2}{x} = 0$

<u>Same</u>

$\frac{1x^2}{2x^2} = \frac{1}{2}$ (1st Coefficient)

<u>Bigger</u>

$\frac{x}{x^2} = \infty\ (DNE)$

Identities (RP)

$\sin = \frac{1}{\csc}$

$\sin^2 + \cos^2 = 1$

$\cos = \frac{1}{\sec}$

$\tan^2 + 1 = \sec^2$

$\tan = \frac{1}{\cot} = \frac{\sin}{\cos}$ $1 + \cot^2 = \csc^2$

Asymptotes (VH)

V : Denominator = 0

H : S = 0 S = 1st Coefficient B = ∞

Intercepts (XY)

x @ y = 0

y @ x = 0

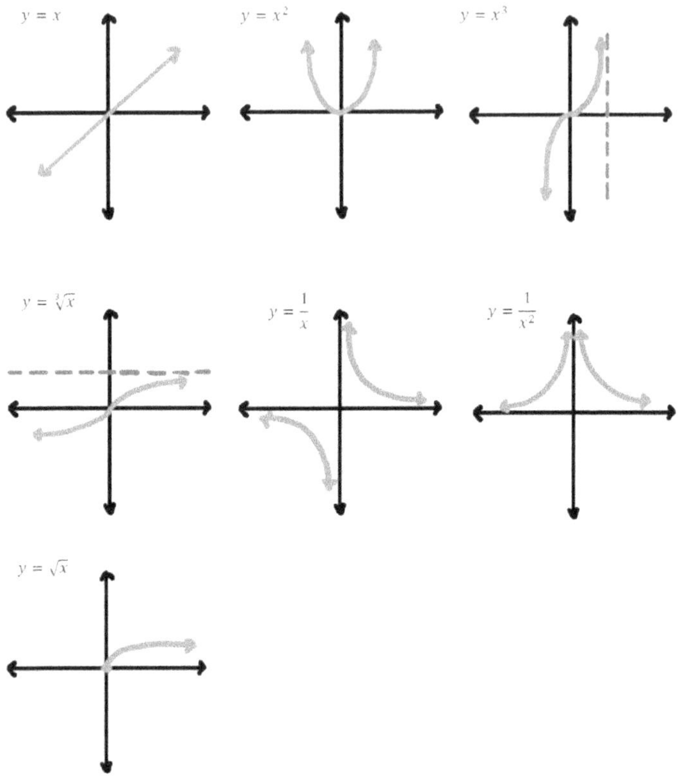

y = x
y = x²
y = x³
y = ∛x
y = 1/x
y = 1/x²
y = √x

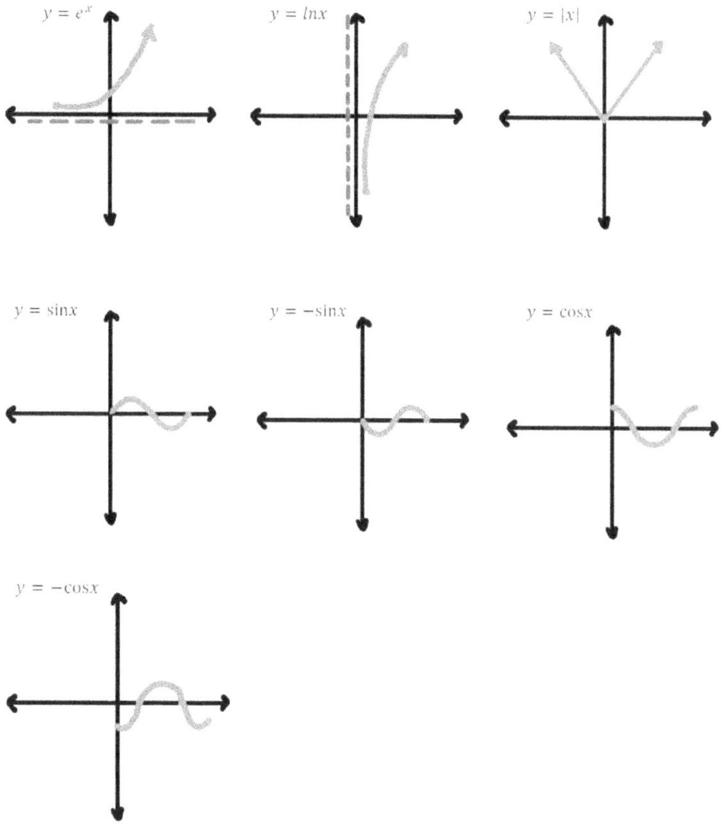
y = e^x
y = lnx
y = |x|
y = sinx
y = −sinx
y = cosx
y = −cosx

$$y = \sec = \frac{1}{\cos}$$

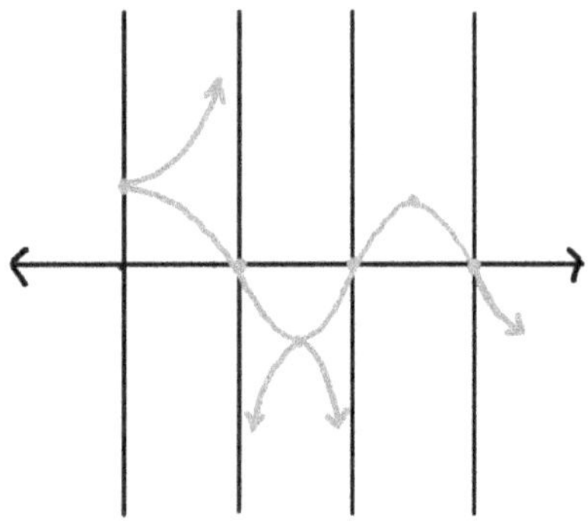

$$y = \csc = \frac{1}{\sin}$$

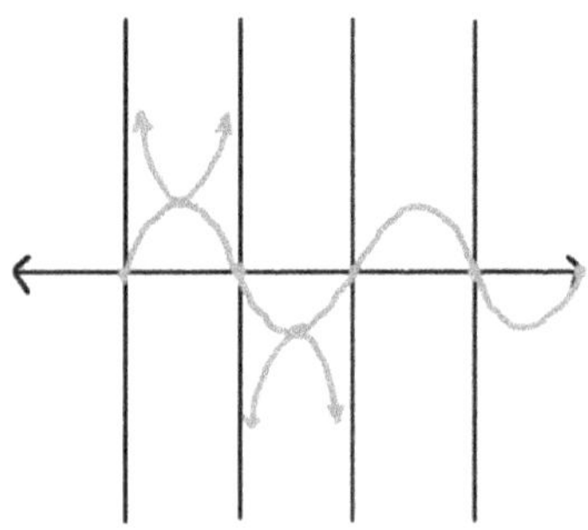

$y = \tan$

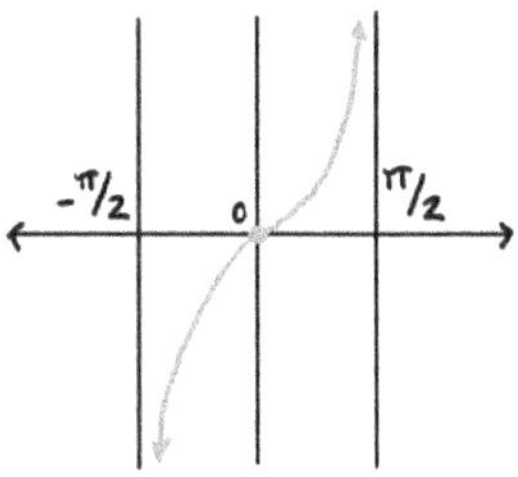

$y = \cot$

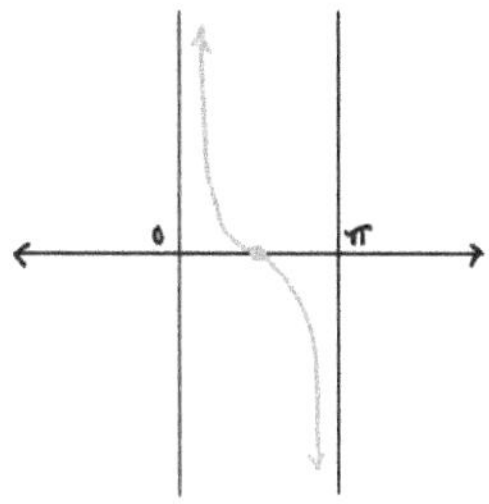

Geometry

Pythagorean Theorem

$$a^2 + b^2 = c^2$$

$$c = \sqrt{a^2 + b^2}$$

Triangles

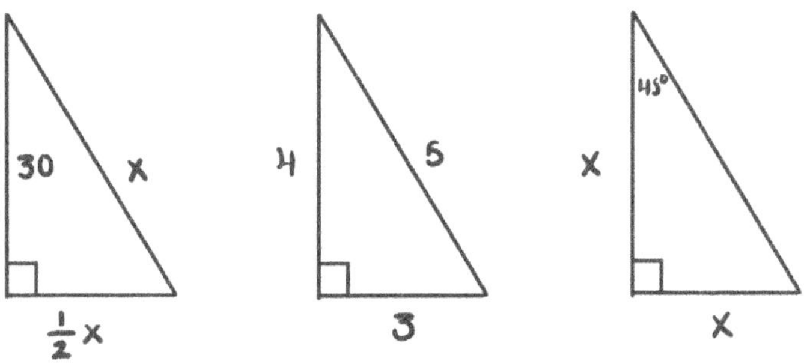

Two Common Proofs

Given

SSS

SAS

ASA

SSA (AC)

HL

CPCTC

Reflexive

Transitive

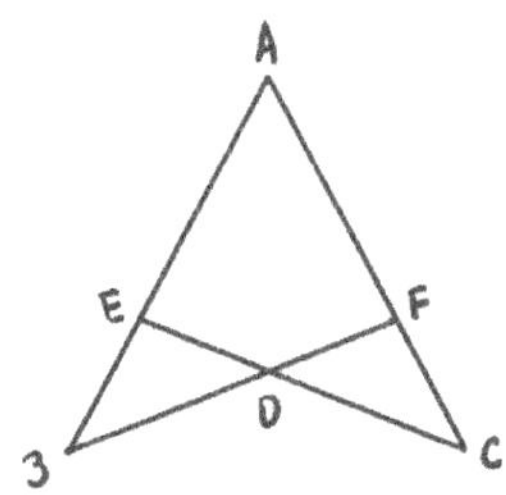

Prove that BF = CE

Two Column Proof

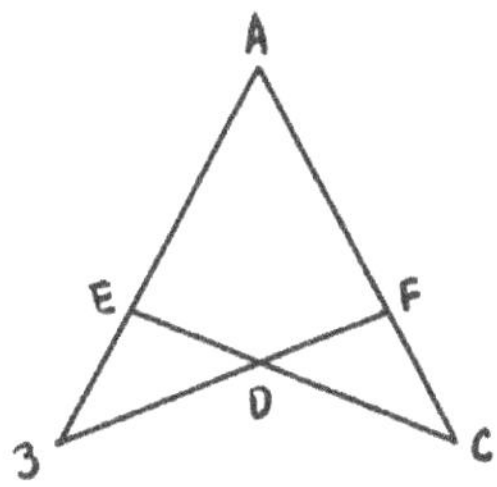

Given

$AB = AC$

$\angle ABF \cong \angle ACE$

Prove

$BF = CF$

Statement	Reason
1.) $AB = AC$	Given
2.) $\angle ABF \cong \angle ACE$	Given
3.) $ABF \cong ACE$	Reflective
4.) $\therefore BF = CE$	CPCTC (Both = AND $\cong$)

Proved

$\sim$ SIMILAR

$\cong$ CONGRUENT ($\therefore$ BY DEFINITION $=$)

Law of Sines

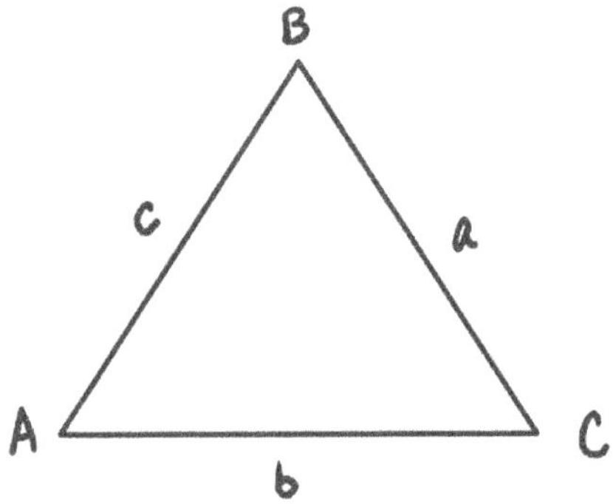

$$\frac{a}{sinA} = \frac{b}{sinB} = \frac{c}{sinC}$$

Ambiguous Case SSA

- $180° - calc + given >$ $\triangle_1$ $180°$ $\therefore$
- $180° - calc + given <$ $\triangle_2$ $180°$ $\therefore$
- If sine < -1 or $> +1 \therefore$ $No\ Solution$

Domain (-1 to +1)

Law of Cosines

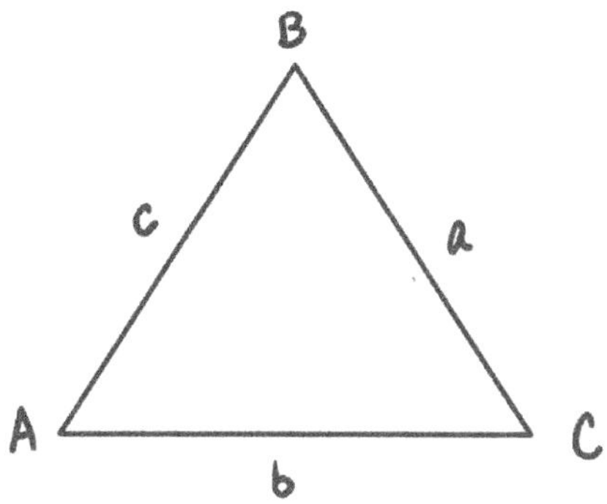

$a^2 = b^2 + c^2 - 2bc\cos A$

Note: Used to solve vectors

Ambiguous Case

$A = 40°$ $a = 8$ $b = 9$

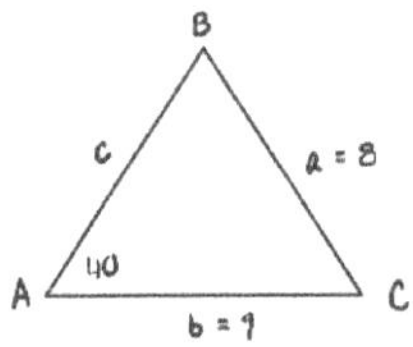

$\left(\frac{a}{\text{sinA}} = \frac{b}{sinB} = \frac{c}{sinc}\right)$ Formula

$\frac{8}{\sin 40} = \frac{9}{sinB}$ $sinB = \frac{9sin40}{8} = 0.7231$

$B = sin^{-1}(0.7231) = 46.3°B$ calc

$180 - 40 + 46.3 = 93.7°C$

$180° - 46.3 = 133.7° + 40 < 180° \therefore$ 

Calc B2 Given

$180° - 40 - 133.7 = 6.3°$

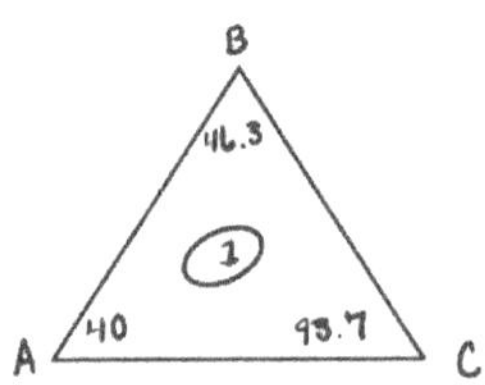

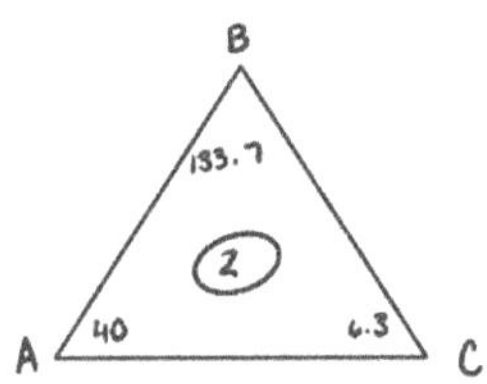

Sequences / Series / Summation

Arithmetic

<u>Sequence</u>

$a_n = a_1 + (n-1)d$

<u>Series</u>

$S_n = \frac{n}{2}(a_1 + a_n)$

Geometric

<u>Sequence</u>

$a_n = a_1 r^{n-1}$

<u>Series</u>

$S_n = \frac{a_1(1-r^h)}{1=r}$ or $\frac{a_1 - a_n{}^r}{1-r}$

(unknown terms)

Summation

$S = \sum_{x=1}^{7} f(x)$

Notes

n = terms

d = $a_2 - a_1$ or $n_2 - n_1$ difference

$r = \frac{a_2}{a_1}$ ratio

Converge

$\frac{1}{2} + \frac{1}{4} + \frac{1}{8} \ldots point$

Diverge

$2 + 4 + 8 \ldots \infty$

Chapter Two

Arithmetic (7)

OOO	**Order of Operation**
^	**Power**
$\sqrt{x}$	**Root**
RI	**Rational/Irrational**
LCD	**Least Common Denominator**
$\|x\|$	**Absolute Value**
$\cdot \rightarrow F$	**Decimal to fraction**

Common Mistakes

There are seven common mistakes that many of my calculus students have made over the last 12 years that I have been teaching them. These are mostly arithmetic errors, not calculus errors. Therefore, I am going to show you an example of the correct solution to each of these.

OOO	**Order of Operation**
^	**Power**
$\sqrt{x}$	**Root**
RI	**Rational/Irrational**
LCD	**Least Common Denominator**
$\lvert x \rvert$	**Absolute Value**
$\cdot\!\to F$	**Decimal to fraction**

Order of Operations

The acronym is called "PEMDAS." It means that you:

Reduce all parentheses and exponents.

Reduce all multiplication and division.

Reduce all addition and subtraction.

Each time that you reduce one of these items, then you return to the left and move to the right of the expression, for example:

$6 \div 2(1 + 2)$

$6 \div 2 \times 3$

3×3

9

Exponents and Roots

Exponents

$x^2 \cdot x^3 = x^{2+3} = x^5$ Add exponents.

$(x^2)^3 = x^{2 \cdot 3} = x^6$ Multiply exponents.

Roots

$\sqrt{625}$

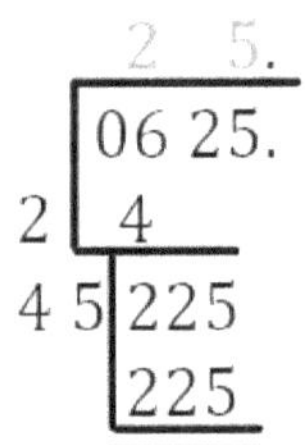

The highest square that goes into 6.

Double the 2 to a 4 guess?

Try 5

No remainder, no rounding necessary.

Rational vs. Irrational Numbers

Rational

$4, \frac{1}{4}$ ends or repeats.

$\left[\frac{x^2}{x^3}\right]$

Irrational

π does not end, does not repeat.

$\left[\frac{x^2}{x^2} \text{ or } \frac{x^3}{x^2}\right]$

Least Common Denominator

$$\frac{1}{2}+\frac{1}{3}+\frac{1}{4}$$

Common Denominator

$$\frac{12+8+6}{24}$$

Least Common Denominator

$$\frac{13}{12}$$

Answer

$$1\frac{1}{12}$$

Always multiply the denominators and then reduce the expression.

Absolute Value

Definition: The distance from zero on a number line.

Rule: There is always a positive and a negative value (both)

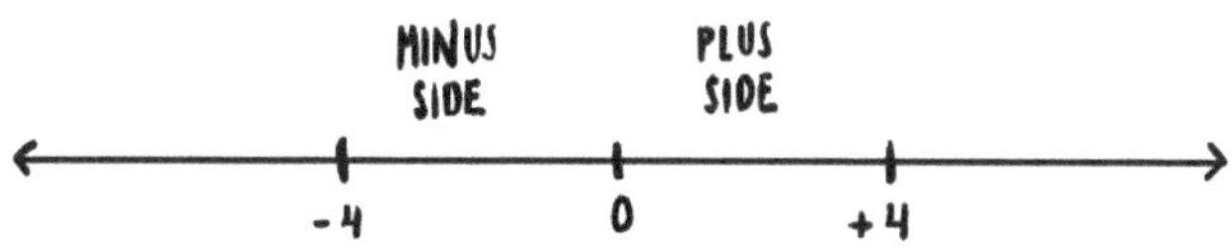

$|x| = |-4| = 4$

$|x| = |4| = 4$

So ± 4 is the answer

Changing a Decimal to a Fraction

$$0.28 \times \frac{100}{100}$$

$$\frac{28}{100}$$

$$\frac{14}{50}$$

$$\frac{7}{25}$$

Chapter Three

Limits

0 (+,-,.) Above, below, at zero

a (+,-,.) Above, below, at any number

∞ **(+,-)** Positive, negative infinity

Try (±0.1) Positive, negative 0.1

S Substitute

F Factor

$\boldsymbol{C_L^R}$ Conjugate (rationalize, least common denominator)

$\boldsymbol{p_u^4}$ Piecewise function (positive, negative, at, function, union)

S Simple

C Complex

C Continuous

I Identities (reciprocal, Pythagorean, double, half, power)

S3 Sine (3 each)

C2 Cosine (2 each)

SSB Smaller, same, bigger

RNC Removable, non-removable, continuous, non-continuous

S Sine

C Cosine

L'Hopital's d/dx numerator, d/dx denominator

$$f(a) = lim\, x \rightarrow (a)$$

If the value of a function of a numerator equals the limit as X approaches that number, then the function is continuous.

Limits definition

Zooming in

Limits Chart 33

0 +,-, . |	S	S	S3	S
a +,-, .	F	C	C2	C
∞ ±	C_L^R	C	SSB	L'H
Try (±, .1)	p_U^4	Identities (Reverse, Pythagorean, double, half, power).	RNC (removable, non-removable, continuous)	$f(a) = \lim_{x\to a}$ continuous

Note:

For an explanation of the symbols, nomenclature, and terminology, refer to the index.

+,-,.

Try ($\pm .1$)

Substitute

Factor

Conjugate

Rationalize

LCD

Piecewise (4)

Union

Identities

Reciprocal

Pythagorean

Double

Half

Power

Sine (3)

Cosine (2)

Smaller

Same

Bigger

Removeable

Non-removeable

Continuous

Non-continuous

Sine

Cosine

L'Hopital's

$f(a) = lim(x) \to (a)$

Continuous

$\neq$ non-continuous

Problems (8)

$\lim_{x \to 2} \frac{x^2 - 4}{x - 2}$	Factor
$\lim_{x \to 5} x^2 + 2x - 4$	Sub
$\lim_{x \to 3} \frac{x^3 - 27}{x - 3}$	Factor
$\lim_{x \to 3} \frac{\frac{1}{x} - \frac{1}{3}}{x - 3}$	LCD
$\lim_{x \to 9} \frac{\sqrt{x} - 3}{x - 9}$	Conjugate
$\lim_{x \to 4} \frac{\frac{1}{\sqrt{x}} - \frac{1}{2}}{x - 4}$	LCD, Conjugate and Substitute
$\lim_{x \to 3} \frac{x^2 - 9}{x - 3}$	Function fractions, try L'Hopital's.
$[-\infty, -3) \cup [+4, +\infty]$	Piecewise f(x) where x = (-2)

Graph

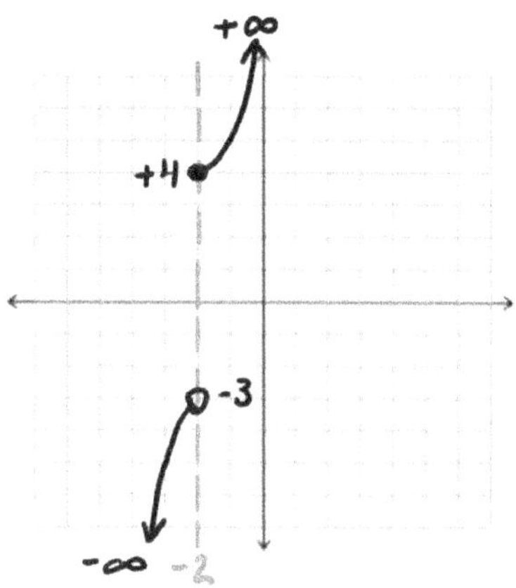

$$\lim_{n \to (-2)^-} = -3$$

$$\lim_{n \to (-2)^+} = +4$$

$$\lim_{n \to (-2)} = DNE$$

f(-2) = +4

$$f(-2) \neq \lim_{n \to (-2)}$$

Not continuous

1. $\lim_{x \to 2} \frac{x^2-4}{x-2}$

Solve by factoring.

$$\frac{\cancel{(x-2)}(x+2)}{\cancel{(x-2)}}$$

$$x + 2$$

$$2 + 2$$

$$4$$

2. $\lim_{x \to 5} x^2 + 2x - 4$

Solve using substitution.

$5^2 + 2(5) - 4$

$25 + 10 - 4$

31

3. $\lim_{x \to 3} \frac{x^3 - 27}{x - 3}$

Solve by factoring.

$$\frac{\cancel{(x-3)}(x^2+3x+9)}{\cancel{(x-3)}}$$

$$x^2 + 3x + 9$$

$$3^2 + 3(3) + 9$$

$$9 + 9 + 9$$

$$27$$

4. $\lim_{x \to 3} \frac{\frac{1}{x} - \frac{1}{3}}{x - 3}$

Solve using the least common denominator.

$$\frac{(3 - x)}{3x} \cdot \frac{1}{x - 3} \cdot \frac{-1}{-1}$$

$$\frac{-1}{3x}$$

$$\frac{-1}{3(3)}$$

$$-\frac{1}{9}$$

5. $\lim_{x \to 9} \frac{\sqrt{x}-3}{x-9}$

Solve using conjugation.

$$\frac{\sqrt{x}-3}{x-9} \cdot \frac{\sqrt{x}+3}{\sqrt{x}+3}$$

$$\frac{\cancel{(x-9)}}{\cancel{(x-9)}\left(\sqrt{x}+3\right)}$$

$$\frac{1}{\sqrt{x}+3}$$

$$\frac{1}{3+3}$$

$$\frac{1}{6}$$

6. $\lim_{x \to 4} \frac{\frac{1}{\sqrt{x}} - \frac{1}{2}}{x-4}$

Solve using the least common denominator,

conjugation, and substitution.

$$\frac{2-\sqrt{x}}{2\sqrt{x}} \cdot \frac{1}{(x-4)} \cdot \frac{2+\sqrt{x}}{2+\sqrt{x}}$$

$$\frac{(4-x)}{(x-4)\left(2\sqrt{x}\right)\left(2+\sqrt{x}\right)} \cdot \frac{-1}{-1}$$

$$\frac{-1}{2\sqrt{x}\left(2+\sqrt{x}\right)}$$

$$\frac{-1}{2\sqrt{4}\left(2+\sqrt{4}\right)}$$

$$\frac{-1}{4(4)}$$

$$\frac{-1}{16}$$

7. $\lim_{x \to 3} \frac{x^2-9}{x-3}$

Solve using function fractions and L'Hopital's rule.

$$\frac{\cancel{(x-3)}(x+3)}{\cancel{(x-3)}}$$

$x + 3$

$3 + 3$

6

8. $[-\infty, -3) \cup [+4, +\infty]$

Piecewise (4).

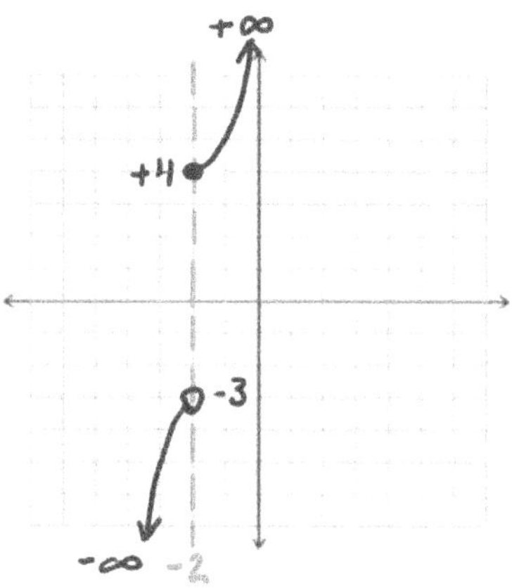

$\lim_{n \to (-2)^-} = -3$

$\lim_{n \to (-2)^+} = +4$

$\lim_{n \to (-2)} = DNE$

f(-2) = +4

$f(-2) \neq \lim_{n \to (-2)}$

Not continuous

1. Factor.

$$\lim_{x \to 3} \frac{x^2 - 9}{x - 3}$$

$$\frac{(x+3)\cancel{(x-3)}}{\cancel{(x-3)}}$$

$$x + 3$$

$$3 + 3$$

$$6$$

2. Try ±.1.

$$\frac{x^2 - 9}{x - 3}$$

$$\frac{(3.1)^2 - 9}{3.1 - 3}$$

$$\frac{9.61 - 9}{.1}$$

$$\frac{.61}{.1}$$

$$6.1$$

$$\approx 6$$

3. L'Hopital's.

$$\frac{x^2 - 9}{x - 3} \cdot \frac{d / dx}{d / dx}$$

$$\frac{2x}{1}$$

$$2(3)$$

$$6$$

4. Conjugate.

$$\frac{x^2 - 9}{x - 3} \cdot \frac{x^2 + 9}{x^2 + 9}$$

$$\frac{x^4 - 81}{(x - 3)(x^2 + 9)}$$

$$\frac{\cancel{(x^2 + 9)}(x^2 - 9)}{(x - 3)\cancel{(x^2 + 9)}}$$

$$\frac{(x + 3)\cancel{(x - 3)}}{\cancel{(x - 3)}}$$

$$x + 3$$

$$3 + 3$$

$$6$$

Quiz

Smaller same bigger (SSB)

$$\lim_{x \to (a)}$$

$\frac{x^2}{x^3}$ $\frac{smaller}{bigger} = 0$ (rational)

$\frac{x^2}{x^2}$ $\frac{same}{same} = 1^{st}$ coefficient (irrational)

$\frac{x^3}{x^2}$ $\frac{bigger}{smaller} = \infty$ (irrational)

Note:

Use polynomial division to change from irrational to rational.

1. $\lim_{x \to \infty} \frac{x^3+5x}{2x^3-x^2+4}$

Say: $\frac{x^3}{2x^3}$

Both variables have the same exponential power,

so we take the first coefficient from each.

The solution is $\frac{1}{2}$

2. $\lim_{x \to 0} \sqrt{9x^2 + x} - 3x$

Use L'Hopital's rule to solve.

Take the derivative of $(9x^2 + x)^{\frac{1}{2}} - 3x$.

$$\frac{1}{2}(9x^2 + x)^{\frac{-1}{2}} \cdot (18x + 1) - 3$$

$$\frac{18x + 1}{2\sqrt{9x^2 + x}} - 3$$

$$\frac{18x}{6x} - 3$$

$$0 - 3$$

-3

Sine 3

$$\frac{\sin x}{x}, \frac{x}{\sin x}, \frac{\sin 5x}{5x}$$

1

Cosine 2

$cos\, x, cos\, 5x$

1

Sine

$$\frac{sin\ 2x}{7x}$$

$$\frac{2}{7} \cdot 1$$

$$\frac{2}{7}$$

Cosine 1.

$$\frac{\cos 3x - \cos 5x}{x^2}$$

$$\frac{5^2 - 3^2}{2}$$

$$\frac{25 - 9}{2}$$

$$\frac{16}{2}$$

$$8$$

Chapter Four

Differentiation

$^{dy}/_{dx}$ Rules (10)

Differentiation

DD C P M S D P Q C I

DD	**Differentiation by definition**
C	**Constant**
P	**Power**
M	**Multiple**
S	**Sum**
D	**Difference**
P	**Product**
Q	**Quotient**
C	**Chain**
I	**Implicit**

Definition: Rate of Change

Differentiation

$\dot{D}\ D_2\ C\ \dot{P}\ m\ s\ D\ \dot{P}\ \dot{Q}\ \dot{C}\ \dot{I}$

$$\frac{f(x+h)-f(x)}{lim\, h \to 0 \quad h}$$

DD – Differentiation by Definition

nx^{n-1}

P – Power

$u'v + u\,v'$

P – Product

$u'\,v\ -u\,v'/v^2$

Q – Quotient

$f'\big(g'(x')\big)$

C – Chain

$x'\,dx/dx + y'\,dy/dx$

Differentiation problems

1. Definition

$d/dx = f(x^2)$

$$\lim h \to 0 \qquad \frac{(x+h)^2 - x^2}{h}$$

$$\frac{x^2 + 2xh + h^2 - x^2}{h}$$

$$\frac{\cancel{h}(2x+h)}{\cancel{h}}$$

$$2x + \cancel{h}$$

$$2x$$

$$\boldsymbol{{}^{d}/_{dx}}$$

2.) Definition $\lim h \to 0 \quad \frac{f(x+h)-f(x)}{h}$

$${}^{d}/_{dx} f(\sqrt{x})$$

$$\lim h \to 0 \quad \frac{\sqrt{x+h}-\sqrt{x}}{h}$$

$$\frac{\cancel{x}+\cancel{h}\cancel{-x}}{\cancel{h}(\sqrt{x+h}+\sqrt{x})}$$

$$\frac{1}{\sqrt{x+\cancel{h}}+\sqrt{x}}$$

$$\frac{1}{\sqrt{x}+\sqrt{x}}$$

$$\frac{1}{2\sqrt{x}}$$

$\boldsymbol{d/_{dx}}$

3.) Power

nx^{n-1}

$d/_{dx}\,(6x^3 + 5x^2 + 9)$

$18x^2 + 10$

$$\boldsymbol{{}^{d}\!/_{dx}}$$

4.) <u>Product</u>

$u'v + uv'$

${}^{d}\!/_{dx}\,(x^3 \sin x)$

$3x^2 \sin x + x^3 \cos x$

$x^2(3 \sin x + x \cos x)$

$^{d}/_{dx}$

5.) <u>Quotient</u>

$(u'v - uv')/v^2$

Problem $^{d}/_{dx}\left[\frac{\sin x}{x^4}\right]\begin{matrix} u \\ v \end{matrix}$

$\left[\frac{x^4 \cos x - 4x^3 \sin x}{x^8}\right]$

$\frac{x^3(x \cos x - 4 \sin x)}{x^8}$

$\frac{x \cos x - 4 \sin x}{x^5}$

$$\mathbf{{}^{d}/_{dx}}$$

6.) <u>Chain</u> $f'(g'(x'))$

$${}^{d}/_{dx}\left[\frac{1}{15}(3x-2)^5\right]$$

$$\frac{5}{15}(3x-2)^4 \cdot 3$$

$$\frac{15}{15}(3x-2)^4$$

$$(3x-2)^4$$

$$\mathbf{{}^{d}/_{dx}}$$

7.) Implicit $\quad x'^{dx}/_{dx} + y'^{dy}/_{dx}$

$${}^{d}/_{dx}\,[x^2y^3 - xy - 10]$$

$$2x\frac{dx}{dx}y^3 + 3y^2\frac{dy}{dx}x^2 - \left[\frac{dx}{dx}y + \frac{dy}{dx}x\right]$$

$$2xy^3 + x^2 3y^2\frac{dy}{dx} - y - x\,{}^{dy}/_{dx}$$

$$(3x^2y^2 - x)\,{}^{dy}/_{dx} = y - 2xy^3$$

$${}^{dy}/_{dx} = \frac{y-2xy^3}{3x^2y^2-x}\cdot\frac{-1}{-1}$$

$$\mathbf{{}^{dy}/_{dx}} = \frac{2xy^3-y}{x-3x^2y^2}$$

Chapter Five

R **Reverse**

G **Guess and Check**

US **U-Substitution**

P **Parts**

TI **Trig-Identities**

TS **Trig-Substitution**

PF **Partial Function**

<u>SISF</u>

S **Substitute**

I **Identities**

S **SOH CAH TOA**

F **Function**

<u>CRF (PSD) LQ SG</u> u/y'

CRF **Complex Rational Function**

PSD **Polynomial Synthetic Division**

L **Lineal**

Q **Quadratic**

S **Substitution**

G **Group**

U **U-Substitution (Numerator)**

Y' **First Differential (Denominator)**

Chapter Five

Integration

$$\int_a^b Algorithms\ (7)$$

Integration $\int_a^b$

R G **US** **P**$_2$ Ti **Ts** **PF**$_3$

R	Reverse
G	Guess and Check
US	U-Substitution
P	Parts
Ti	Trigonometric Substitution
PF	Partial Fractions

Integration $\int_a^b$

Definition (1)

Quantity (Area & Volume)

Polynomial Division

How to change an irrational expression to rational, then integrate.

$\int \frac{3x^3+2x-11}{x-3} dx$ <u>Note</u>: $\frac{x^3}{x}$: Irrational Expression

$$3x^2 + 9x + 29 + \frac{76}{x-3}$$: Rational Expression

$$x - 3 \overline{) \, 3x^3 + 0x^2 + 2x - 11}$$

$$\underline{3x^3 - 9x^2}$$

$$9x^2 + 2x$$

$$\underline{9x^2 - 27x}$$

$$29x - 11$$

$$\underline{29x - 87}$$

$$+76$$

$$\int 3x^2 dx + \int 9x dx + \int 29 dx + \int \frac{76}{x-3} dx$$

<u>Use Reverse</u>

$\frac{3x^3}{3} + \frac{9x^2}{2} + \frac{29x}{1} + \int \frac{76}{x-3} dx$ (<u>U-Sub</u>)

<u>U-Sub</u>

$$76\int \frac{1}{x-3}dx$$

U=x-3 du=dx dx=du

$$76\int \frac{1}{u}du$$

$$76ln|u| + c$$

$$76ln|x-3| + c$$

$$x^3 + \frac{9}{2}x^2 + 29x + 76ln|x-3| + c$$

$$\int_a^b$$

<u>Algorithms</u> (7)

R G **US** $\mathbf{P_2}$ Ti **Ts** $\mathbf{PF_3}$

1.) $\frac{X^{N+1}}{N+1}$ Reverse

2.) U = I du = I'dx dx = du/I' U-Sub.

3.) $uv - \int v\,du$

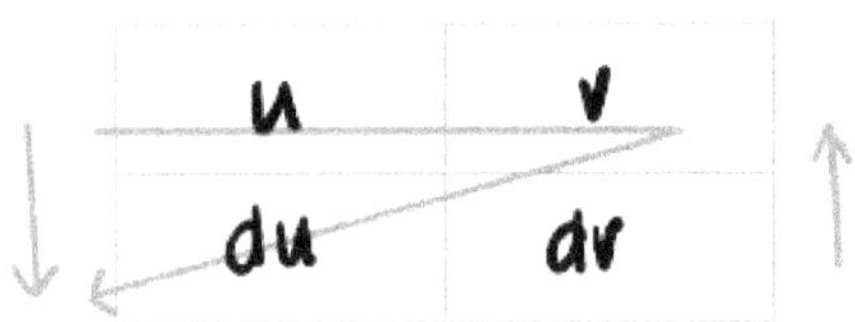

Parts_2

4.) <u>SISF</u> $\sqrt{a^2 - x^2}$ x = a$\sin\theta$ T-Sub.

$\sqrt{a^2 + x^2}$ x = a$\tan\theta$

$\sqrt{x^2 - a^2}$ x = a$\sec\theta$

5.) <u>CRF(PD) LQ   SG   U/y'</u>

$\frac{A}{x+1}$ $\frac{Ax+B}{x^2+1}$

P – Fraction$_3$

$$\frac{Ax+B}{x^2+1}+\frac{Cx+D}{(x^2+1)^2}$$

$$\int_a^b$$

Problems (8)

1.) Reverse

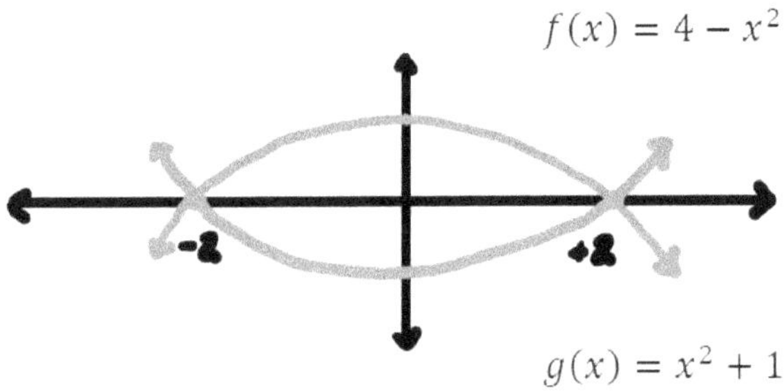

$$A = \int_{-2}^{+2}[f(x) - g(x)]dx$$

$$\int_{-2}^{+2}[(4 - x^2) - (x^2 + 1)]dx$$

$$\int_{-2}^{+2}(-2x^2 + 3)dx = \left[\frac{-2x^3}{3} + 3x\right]_{-2}^{+2}$$

$$\left[\frac{-2(2)^3}{3} + 3(2)\right] - \left[\frac{-2(-2)^3}{3} + 3(-2)\right]$$

$$-\frac{16}{3} + 6 - \frac{16}{3} + 6 = -\frac{32}{3} + \frac{36}{3} = 4/3$$

$$\int_a^b$$

2.) U-sub

$$\int (3x-2)^4 dx$$

$u = (3x - 2)$ du = 3dx dx

= du/3

$$\int u^4 \frac{du}{3}$$

$$\frac{1}{3} \cdot \frac{u^5}{5}$$

$$\frac{1}{15} \cdot u^5$$

$$\frac{1}{15}(3x-2)^5 + c$$

U-Substitution Extra

$$\int [(1 + tanx^7)sec^2x]dx$$

$$u = (1 + tansx) \quad du = sec^2xdx \quad dx = \frac{du}{sec^2x}$$

$$\int u^7 \cancel{sec^2x} \frac{du}{\cancel{sec^2x}}$$

$$\int u^7 du = \frac{u^8}{8} + c$$

$$\frac{1}{8}(1 + tanx)^8 + c$$

U-Sub Extra

$$\int \sec(5\theta)\tan(5\theta)\,d\theta$$

$u = 5\theta$ $\qquad du=5d\,\theta$ $\qquad d\,\theta = \frac{du}{5}$

$$\int \sec(u)\tan(u)\frac{du}{5}$$

$$\frac{1}{5}\int (secutanu)du$$

$$\frac{1}{5}secu + c$$

$$\frac{1}{5}\sec(5\theta) + c$$

$$\int_a^b$$

3.) <u>Parts</u> (1 STEP)

$$\int \underset{u}{x}\,\underset{dv}{e^x\, dx}$$

$$uv - \int v du$$

u	v
x	e^x
dx	$e^x dx$
du	dv

$$xe^x - \int e^x dx$$

$$xe^x - e^x + c$$

4.) <u>Parts</u> (2 Step)

$\int x^2 sinxdx$

1.)

u	v
x^2	$-cosx$
$2xdx$	$sinxdx$
du	dv

$$-x^2cosx - \int -cosx2xdx$$

$$-x^2cosx + \left[\int 2xcosdx\right]$$

2.)

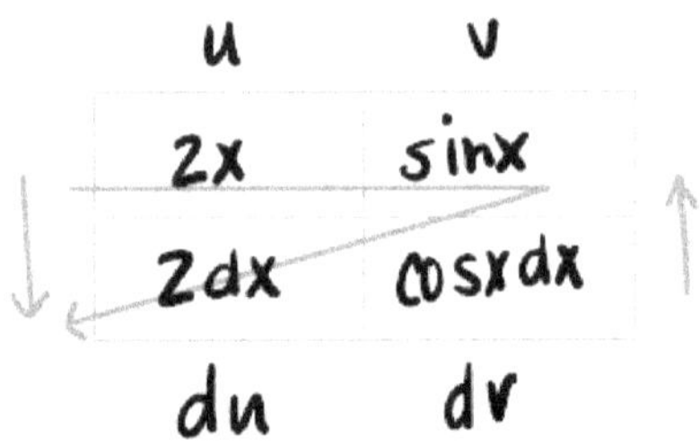

$2xsinx - \int sinx2dx$

$2xsinx - (-2cosx)$

$[2xsinx + 2cosx]$

$-x^2cosx + [2cosx + 2xsinx]$

$-x^2cosx + 2cosx + 2xsinx + c$

$$\int_a^b$$

<u>Parts</u> (Special) $uv - \int vdu$

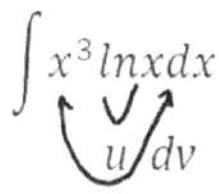

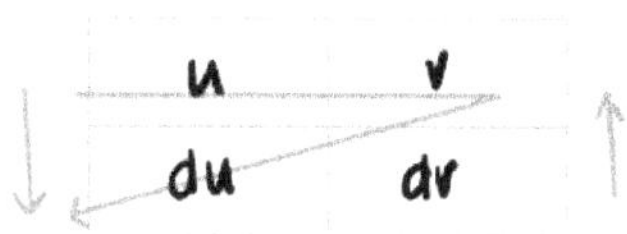

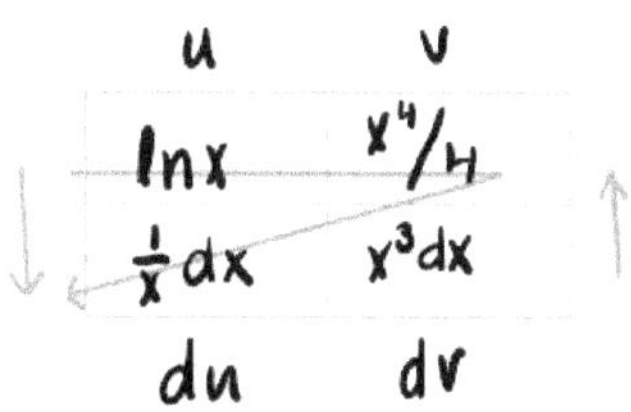

$$\frac{x^4}{4} lnx - \int \frac{x^3}{4} \cdot \frac{1}{\cancel{x}} dx$$

$$\frac{1}{4} x^4 lnx - \frac{1}{4} \int x^3 dx$$

$$\frac{1}{4} x^4 lnx - \frac{1}{4}\left(\frac{x^4}{4}\right) + c$$

$$\frac{1}{4} x^4 lnx - \frac{1}{16} x^4 + c$$

$\int_a^b$ *Page* 1

5.) T-SUB (SISF)

$\int \frac{\sqrt{4-x^2}}{x^2} dx$

SUB, IDENTITIES,

SHO CAH TOA, FUNCTION

<u>SUB</u>	$\sqrt{a^2 - x^2}$	$x = a \sin \theta$
$a^2 = 4\, a = 2$	$\sqrt{a^2 + x^2}$	$x = \theta \tan \theta$
$x = a \sin \theta = 2 \sin \theta$	$\sqrt{x^2 - a^2}$	$x = a \sec \theta$

$dx = 2 \cos \theta d\theta$

<u>IDENTITY</u>

$$\int \frac{\sqrt{4 - 4 \sin^2 \theta}}{4 \sin^2 \theta} 2 \cos \theta d\theta$$

$$\int \frac{\sqrt{4(1 - \sin^2 \theta)}}{4 \sin^2 \theta} 2 \cos \theta d\theta$$

$$\int \frac{\cancel{2}\sqrt{1 - \sin^2 \theta}}{4 \sin^2 \theta} \cancel{2} \cos \theta d\theta$$

$$\int \frac{\sqrt{\cos^2 \theta}}{\sin^2 \theta} \cos \theta d\theta$$

$\int_a^b$ *Page* 2 (T-SUB)

$$\int \frac{\cos^2\theta}{\sin^2\theta}\,d\theta$$

$$\int \frac{1-\sin^2\theta}{\sin^2\theta}\,d\theta$$

$$\int \left[\frac{1}{\sin^2\theta} - 1\right] d\theta$$

$$\int [\csc^2\theta - 1] d\theta$$

$$-\cot\theta - \theta + c \leftarrow$$

<u>SOH CAH TOA</u>

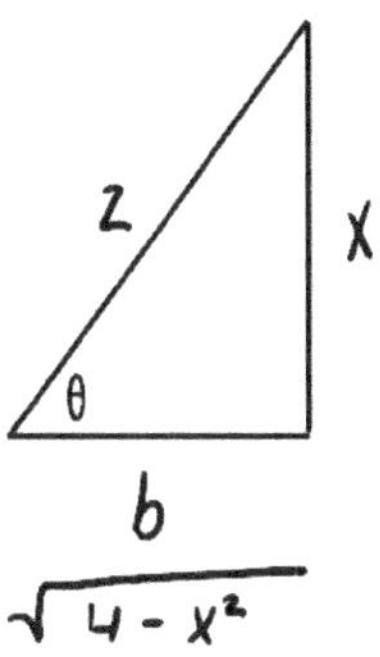

$x = 2\sin\theta$ $\qquad 2^2 = x^2 + b^2$

$\sin\theta = \frac{x}{2}$ $\qquad b^2 = -x^2 + 4$

$\theta = \sin^{-1}\frac{x}{2}$ $\qquad b = \sqrt{4 - x^2}$

<u>FUNCTION</u>

$$-\cot\theta - \theta + c = -\frac{\sqrt{4-x^2}}{x} - \sin^{-1}\frac{x}{2} + c$$

Decomposition $\int_a^b$ *Problem* 1 (*P Fraction*)

6.) $\int \frac{5}{x^2+x-6} dx$ <u>P-Fraction</u>

$(x-2)(x+3)$ Linear CRF (PD) LQ SG $^u/_y{}'$

Complex, Rational,

Function, Polynomial

Division

Linear, Quadratic,

Sub, Group, U-Sub,

1st $^d/_{dx}$

$$(x-2)(x+3)$$

$$\frac{5}{(x-2)(x+3)} = \frac{A}{x-2} + \frac{B}{x+3}$$

$$5 = A(x+3) + B(x-2)$$

<u>Roots</u> $x = 2 : 5 = A(2+3) = 5A \therefore A = 1$

<u>Sub</u> $x = -3 : 5 = B(-3-2) = -5B \therefore B = -1$

<u>Then</u> $\int \frac{5}{x^2+x-6}\,dx = \int \frac{1}{x-2}\,dx - \int \frac{1}{x+3}\,dx$

$$\ln|x-2| - \ln|x+3| + c$$

$$\ln\left|\frac{x-2}{x+3}\right| + c$$

Decomposition $\int_a^b$ *Problem 2 (P Fraction)*

7.) $\frac{x-9}{x^2+3x-10}$ <u>P-Fraction</u>

$(x-2)(x+5)$ CRF (PD) LQ SG $^u/_{y'}$

Linear

FT
-10
-2 5

$[(x-2)(x+5)]$

$$\frac{x-9}{(x-2)(x+5)} = \frac{A}{x-2} + \frac{B}{x+5}$$

<u>Sub</u>

$x - 9 = A(x+5) + B(x-2)$

$x = 2 : 2 - 9 = A(7) : -7 = 7A \therefore A = -1$

$x = -5 : -5 - 9 = B(-5-2) : -14 =$

$B(-7)$ $B = 2$

$$\int \frac{x-9}{x^2+3x-10}dx = \int \frac{-1}{x-2}dx + \int \frac{2}{x+5}dx$$

$2\ln|x+5| - \ln|x-2| + c$

Decomposition $\int_a^b$ ***Problem* 3 (*P Fraction*)**

8.) $\int \frac{x^2-1}{x^3+x} dx$ CRF (PD) LQ SG $^u/_{y'}$

$\frac{x^2-1}{x(x^2+1)} = \frac{A}{x} + \frac{Bx+c}{x^2+1}$ $[x(x^2+1)]$

$x^2 - 1 = A(x^2+1) + (Bx+c)(x)$

<u>Sub</u>

$x = 0: \ -1 = A(1) + 0 \therefore A = -1$

<u>Group</u>

$x^2 - 1 = Ax^2 + A + Bx^2 + Cx$

$x^2 - 1 = (A+B)x^2 + Cx + A$

$x^2 = (A+B)x^2$

$1 = A + B \therefore B = 2$

$0 = Cx \therefore C = 0$

<u>U-Substitution</u>

$u = x^2 + 1$

$du = 2xdx$

$dx = {}^{du}/_{2x}$

$\int \frac{2x}{u} \frac{du}{2x}$

$$\int \frac{1}{u} du$$

$$\ln|u|$$

$u - sub$

$$\ln|x^2 + 1|$$

$\downarrow$

$$\int \frac{x^2 - 1}{x^3 + x} dx = \int \frac{-1}{x} dx + \int \frac{2x}{x^2 + 1} dx$$

$$\ln|x^2 + 1| - \ln|x| + c$$

$$\ln\left|\frac{x^2 + 1}{x}\right| + c$$

CHAPTER SIX

SPECIAL TOPICS

Topics

DC	Double Curves
WS	Weird Shapes
LD	Logarithmic Differentiation
OPT	Optimization
RR	Related Rates
ID	Inverse Differentiation
LA	Lineal Approximation
NM	Newton Method
CS	Curve Sketching

Topics

Definition

Special types of calculus problems that you need to be able to recognize and solve.

Topics

Problems (a)

DC – Double Curves

WS – Weird Shapes

LD – Logarithmic Differentiation

OPT – Optimization

RR – Related Rates

ID – Inverse Differentiation

LA – Linear Approximation

NM – Newton Method

CS – Curve Sketching

Topics

Formulas

DC	$\int[f(x) - g(x)]\,dx$	(TB, RL)
		Top, Bottom
	Right, Left	
WS	$\int a^2[f(x) - g(x)]dx$	(S, D, W)
	$(\pi r^2, \pi\Delta r^2)$	Slice, Disk,
		Washer
LD	${}^{d}/_{dx}\ln(u) = {}^{u'}/_{u}$	
OPT	C, UC, ${}^{d}/_{dx} = 0$	(Extrema)
RR	C, UC, ${}^{d}/_{dt} = 0$	(Time)
ID	${}^{d}/_{dx} f^{-1}(x)$	(Switch X & Y Values)
LA	$f(x) \approx f(a) + (x - a)f'(a)$	
NM	$x_{n+1} = x_n - \frac{f(x_n)}{f'(x_n)}$	

CS y y' y''' Function Data Table

Sign Chart Graph

Intercepts (CP)

Extrema (m/m)

Concavity (POI)

Topic Problems

Double Curves

Top to Bottom <u>TB</u>

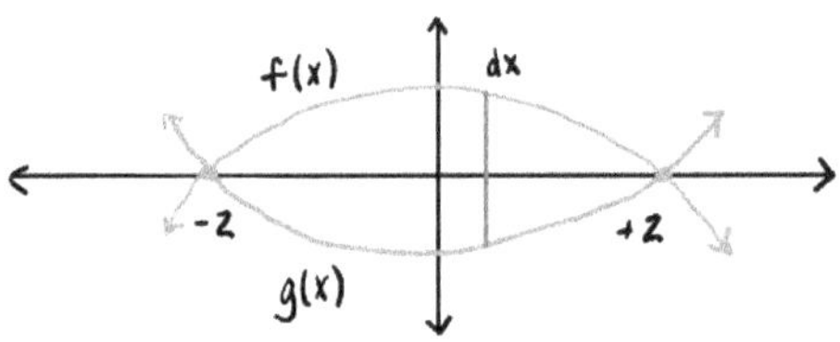

$f(x) = 4 - x^2 \quad g(x) = x^2 + 1 \quad |_{-2}^{+2}$ Limits

$\int_{-2}^{+2} [f(x) - g(x)]dx = \int_{-2}^{+2} [(4 - x^2) - (x^2 + 1)]dx =$

$\left[\frac{-2(2)^3}{3} + 3(2)\right] - \left[\frac{-2(-2)^3}{3} + 3(-2)\right] = -\frac{16}{3} + 6 - \frac{16}{3} +$

$6 = -\frac{32}{3} + \frac{36}{3} = {}^4/_3$

Double Curves Right to Left <u>RL</u> Rotate 90°

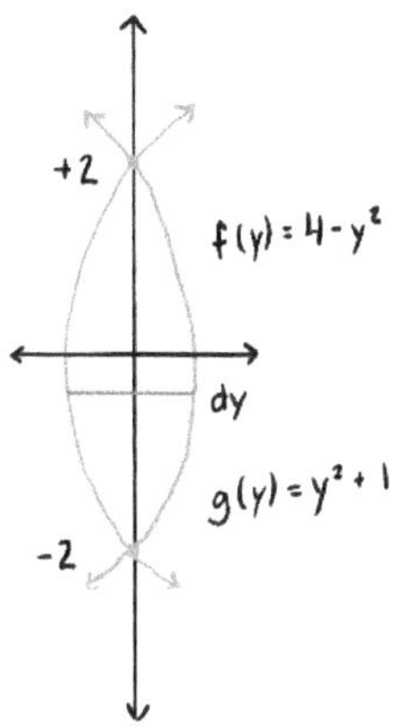

$$\int_{-2}^{+2}[f(y) - g(y)]dy \int_{-2}^{+2}[(4 - y^2) - (y^2 + 1)]\, dy =$$

$See\ Above = {}^{4}/_{3}$

Topic Problems

Find the Volume **Weird Shapes**

<u>Cone</u> <u>Shape</u> <u>Section</u> <u>Ratio</u>

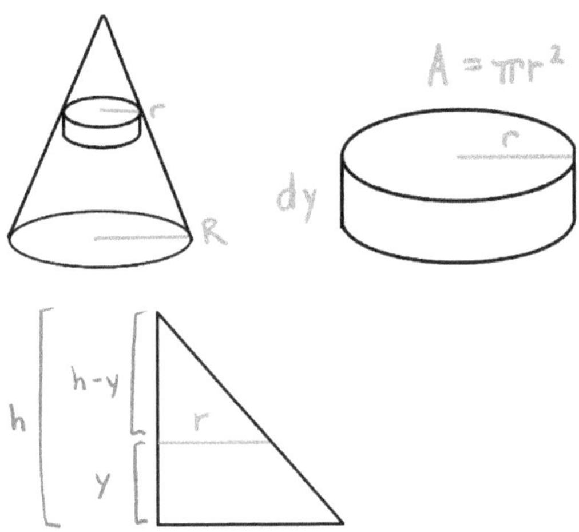

vFormula

$$Volume = \int_0^h \pi r^2 dy = \int_0^h \frac{\pi R^2 (1 - \frac{1}{h})^2}{Area} \frac{dy}{Height}$$

U-Substitute

$$u = (1 - \frac{1}{h} y) \qquad du = \left(-\frac{1}{h}\right) dy$$

$$dy = (-h) du$$

U-Limits

$$u(h) = 0\ Top\ u(0) = 1\ Bottom$$

Solution

$$V = \int_1^0 \pi R^2 u^2 (-h) du = \pi R^2 (-h) \int_1^0 u^2 du =$$

$$\pi R^2 (-h) \left[\frac{u^3}{3}\right]_1^0 = \pi R^2 (-h) \left[0 - \frac{1}{3}\right] = {}^1\!/_3\, \pi R^2 h$$

Topic Problems

Weird Shapes Always have (7) parts: Shape, section, ratio, u-substitution, u-limits, & solution.

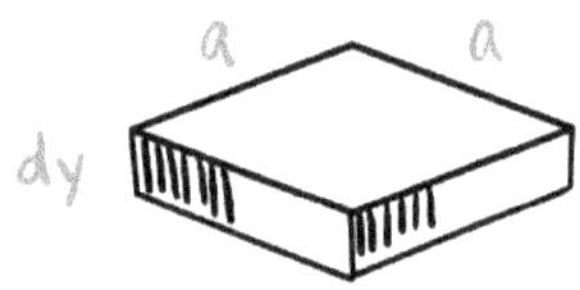

Slice

$\int_a^b a^2[f(y)]dy$

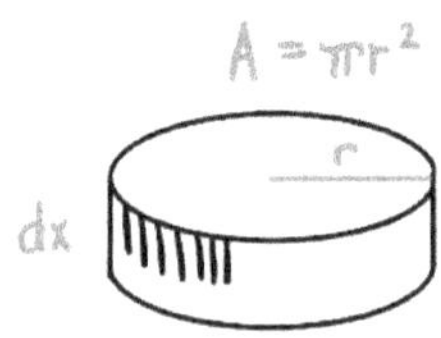

Disk

$\int_a^b \pi r^2[f(x)]dx$

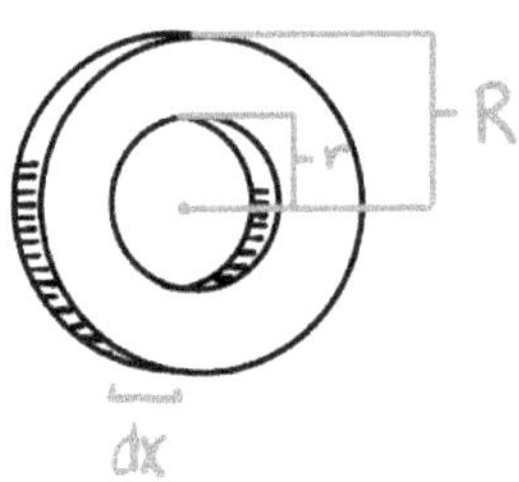

Washer

$\int_a^b \pi\Delta r^2[f(x)]dx \;\; where \;\; \Delta r^2 = (R^2 - r^2)$

Topic Problems

Logarithmic Differentiation

Formula: $^{d}/_{dx} \ln(u) = {}^{u'}/_{u}$

$$y = f(x) = \frac{x^2(6+3x)^4}{\sqrt[3]{9-x^2}}$$

$$\ln y = 2\ln x + 4\ln(6+3x) - \frac{1}{3}\ln(9-x^2)$$

$$\left(\frac{1}{y}\right)\frac{dy}{dx} = \frac{2\left(\frac{1}{x}\right)}{x} + \frac{4(3)}{6+3x} - \frac{-2x}{3(9-x^2)}$$

$$\frac{dy}{dx} = y\left[\frac{2}{x^2} + \frac{12}{6+3x} + \frac{2x}{3(9-x^2)}\right]$$

$$\frac{dy}{dx} = \frac{x^2(6+3x)^4}{\sqrt[3]{9-x^2}}\left[\frac{2}{x^2} + \frac{4}{2+x} + \frac{2x}{3(9-x^2)}\right]$$

Topic Problems

Optimization	Constrained	Unconstrained
(c, UC, Extrema)	First ${}^{d}/_{dx} = 0$	Extrema
	(max, min)	

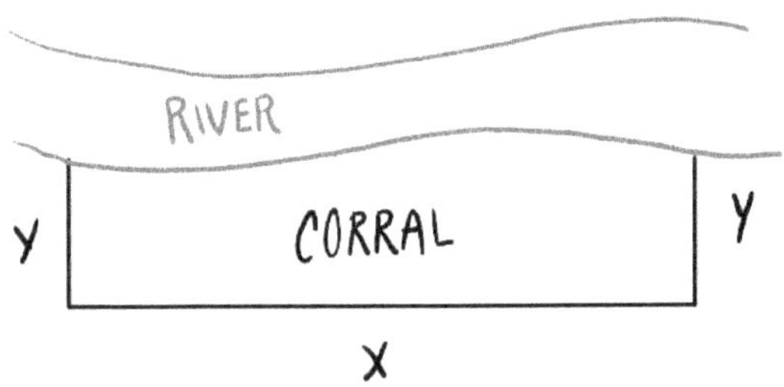

Given: 4000 ft fence

Find: The maximum area the fence will enclose

C: $P = x + 2y$

$4000 = x + 2y$

$\underline{x = 4000 - 2y}$

$A = (4000 - 2y)y$

$4000y - 2y^2$

Find ${}^{d}/_{dx} = 0$

$$4000y = 2y^2$$

$$d/_{dx} = 4000 - 4y$$

$$4y = 4000$$

$$y = 1000$$

$$x = 2y = 2(1000) = 2000$$

<u>UC</u>: $A = xy$

$$A = 2000 \times 1000$$

$$A = 2 \times 10^6\ ft^2$$

$$P = 2000 + 2(1000)$$

$$P = 4000\ ft$$

Topic Problems

Related Rates

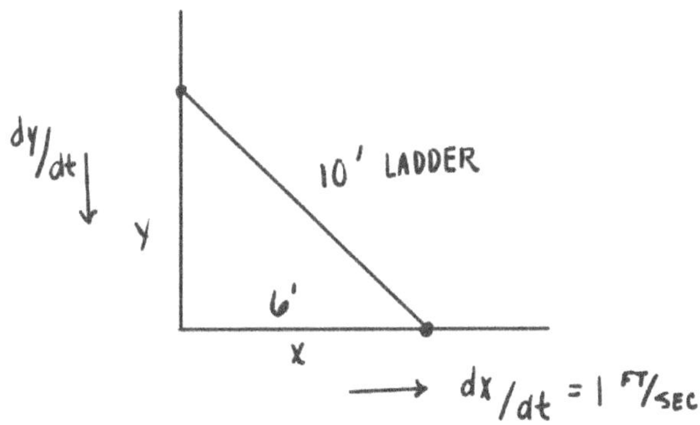

Find: $\frac{dy}{dt} @x = 6'$

$x^2 + y^2 = 10^2 = 100$

$y^2 = 10 - 36 = 64$

$\underline{y = 8}$

First $d/_{dt} = 0$

$d/_{dt} = 2x\, {}^{dx}/_{dt} + 2y\, {}^{dy}/_{dt} = 0$

$2y\, {}^{dy}/_{dt} = -2x\, {}^{dx}/_{dt}$

$$ {}^{dy}/_{dt} = \frac{-2x}{2y} {}^{dx}/_{dt} $$

$$ {}^{dy}/_{dt} = -\frac{6}{8}(1) $$

$$ {}^{dy}/_{dt} = -{}^{3}/_{d}\ {}^{ft}/_{sec} $$ Down

Topic Problems

Inverse Differentiation

${}^{d}/_{dx} f^{-1}(x)$ (Switch x & y)

$f(x) = x^2 - 7$

Find: ${}^{d}/_{dx} f^{-1}(x)$

$y = x^2 - 7$

$x = y^2 - 7$

$y^2 = x + 7$

$y = \sqrt{x+7}$

$f^{-1}(x) = \sqrt{x+7}$

$(f^{-1})'(x) = (x+7)^{\frac{1}{2}} = \frac{1}{2}(x+7)^{-\frac{1}{2}} \cdot 1$

$= \frac{1}{2(x+7)^{\frac{1}{2}}} = \frac{1}{2\sqrt{x+7}}$

Topics Problems

Linear Approximation <u>Arithmetic</u>

$f(x) \approx f(a) + (x-a)f'(a)$ Find $\sqrt[3]{64}$ 63.7

Actual Guess Actual Guess DifferentialGuess Guess

$f(x) \approx \sqrt[3]{64} + (63.7 - 64)f'(\sqrt[3]{64})$

$f'(\sqrt[3]{64})$

$f'(64)^{1/3}$

$\frac{1}{3}(64)^{-2/3}$

$\frac{1}{3(64)^{\frac{2}{3}}}$

$\frac{1}{3(16)}$

$\frac{1}{48}$

$f(\sqrt[3]{63.7}) \approx \sqrt[3]{64} + (-.3)\frac{1}{48}$

$\approx 4 + \frac{-.3}{48}$

$$\approx 4 - (0.00625)$$

$$\approx 3.99375$$

$$\approx 3.9938$$

Topic Problems

Newton Method <u>Algebra</u>

$$x_{n+1} = x_n - \frac{f(x_n)}{f'(x_n)}$$

Find: $f(x) = x^2 - 5$ at [2,3] to 4 decimal places

$2.5 = \frac{5}{2} = x_n$

Start

$x_0 = \frac{5}{2}$ n=0 2.5

$$x_1 = \frac{5}{2} - \frac{\left(\frac{5}{2}\right)^2 - 5}{2\left(\frac{5}{2}\right)} = \frac{9}{4} = 2.25$$

$$x_2 = \frac{9}{4} - \frac{\left(\frac{9}{4}\right)^2 - 5}{2\left(\frac{9}{4}\right)} = \frac{161}{72} = 2.2361$$

$x_3 = \frac{161}{72} - \frac{\left(\frac{161}{72}\right)^2 - 5}{2\left(\frac{161}{72}\right)} = 2.23616$ repeats

$= 2.2361$

Topic Problems

Curve Sketching

y y' y''	Data Table, Sign Chart, Graph
$y = f(x) = x^3 + 6x^2 + 9x$	Intercepts I
$y' = 3x^2 + 12x + 9$	Extrema CP
$y'' = 6x + 12$	Concavity POI

$y = x^3 + 6x^2 + 9x = 0$

$x(x + 3)(x + 3) = 0$

$x = 0, -3$

	X	Y		
I	0	0		−
I, CP	-3	0		MAX
CP	-1	-4	•	MIN
POI	-2	-2	•	+

$*f(-1) = -1^3 + 6(-1)^2 + 9(-1) = -1 + 6 - 9 =$
-4

$*f(-2) = -2^3 + 6(-2)^2 + 9(-2) = -8 + 24 - 18 =$
-2

$y' = 3x^2 + 12x + 9 = 0$

$3(x + 3)(x + 1)$

$x = -3, -1$ CP

$y'' = 6x + 12 = 0$

$6x = -12$

$6x = -2$ POI

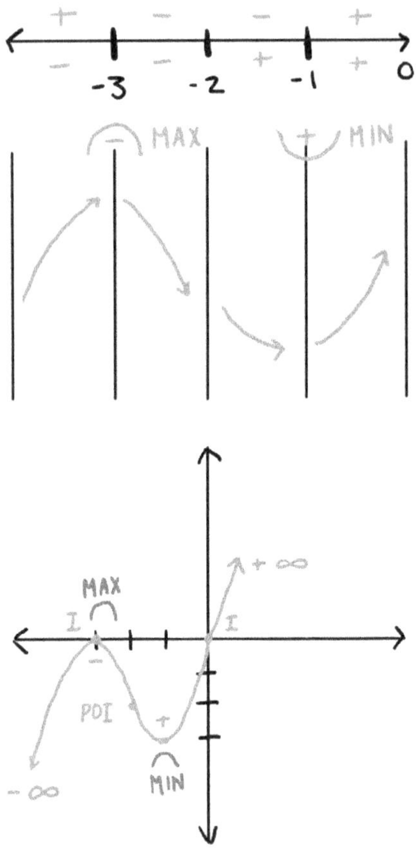
-3
-2
-1
0
MAX
MIN
+∞
MAX
I
I
POI
MIN
-∞

Chapter Seven

Special Problems

Chapter 7

L8 Limits (8)

D7 Differentials (7)

I5 Integrals (5)

A2 Areas (2)

V3 Volumes (3)

S4 Sequences (4)

S4 Series (4)

S2 Summations (2)

Problems

L8 D7 I5 A2 V3

Limits (8 each)

Differentiation (7 each)

Integration (5 each)

Area (2 each)

Volume (3 each)

Sequence (4 each)

Series (4 each)

Summation (2 each)

Problems 25

Definition

Common calculus problems you need to be able to identify and solve.

(5) Categories

8 limits

7 Differentials

5 Integrals

2 Area

3 Volume

Problems

Limits 8

1.) $\lim_{x \to 2} \frac{x^2-4}{x-2}$ Factor.

2.) $\lim_{x \to 5} x^2 + 2x - 4$ Substitute.

3.) $\lim_{x \to 3} \frac{x^3-27}{x-3}$ Factor.

4.) $\lim_{x \to 3} \frac{\frac{1}{x}-\frac{1}{3}}{x-3}$ LCD.

5.) $\lim_{x \to 9} \frac{\sqrt{X}-3}{x-9}$ Conjugate.

6.) $\lim_{x \to 4} \frac{\frac{1}{\sqrt{x}}-\frac{1}{2}}{x-4}$ LCD, Conjugate, Substitute.

7.) $\lim_{x \to 3} \frac{x^2-9}{x-3}$ Function, Factor, Conjugate, Try, L'Hopital's.

8.) $[-\infty, -3) \cup [+4, +\infty]$ Peicewise.
f(x) where *x=(-2)*

Problems Limits 8

1.) $\lim_{x \to 2} \frac{x^2-4}{x-2}$ Factor.

$$\frac{\cancel{(x-2)}(x+2)}{\cancel{(x-2)}}$$

$x + 2$

$2 + 2$

4

2.) $\lim_{x \to 5} x^2 + 2x - 4$ Substitute.

$x^2 + 2x - 4$

$5^2 + 2(5) - 4$

$25 + 10 - 4$

31

3.) $\lim_{x \to 3} \frac{x^3 - 27}{x - 3}$ Factor.

$x^3 - y^3 = (x - y)(x^2 + xy + y^2)$

$$\frac{\cancel{(x-3)}(x^2 + 3x + 9)}{\cancel{(x-3)}}$$

$x^2 + 3x + 9$

$3^2 + 3(3) + 9$

$9 + 9 + 9$

27

Problems

4.) $\lim_{x\to 3} \frac{\frac{1}{x}-\frac{1}{3}}{x-3}$ LCD.

$$\frac{\frac{3-x}{3x}}{\frac{x-3}{1}} \cdot \frac{-1}{-1}$$

$$-\frac{\cancel{(x-3)}}{3x} \cdot \frac{1}{\cancel{(x-3)}}$$

$$\frac{-1}{3x}$$

$$\frac{-1}{3(3)}$$

$$-\frac{1}{9}$$

5.) $\lim_{x \to 9} \frac{\sqrt{X}-3}{x-9}$ Conjugate.

$$\frac{\sqrt{x}-3}{x-9} \cdot \frac{\sqrt{x}+3}{\sqrt{x}+3}$$

$$\frac{(x-9)}{(x-9)\left(\sqrt{x}+3\right)}$$

$$\frac{1}{\sqrt{x}+3}$$

$$\frac{1}{3+3}$$

$$\frac{1}{6}$$

6.) $\lim_{x \to 4} \frac{\frac{1}{\sqrt{x}} - \frac{1}{2}}{x-4}$ LCD, Conjugate, Substitute.

$$\frac{\frac{2-\sqrt{x}}{2\sqrt{x}}}{\frac{x-4}{1}} \cdot \frac{2+\sqrt{x}}{2+\sqrt{x}}$$

$$\frac{\frac{4-x}{2\sqrt{x}(2+\sqrt{x})}}{x-4}$$

$$\frac{\cancel{(4-x)}-1}{2\sqrt{x}(2+\sqrt{x})} \cdot \frac{1}{\cancel{(x-4)}} \cdot \frac{-1}{\cancel{-1}}$$

$$\frac{-1}{4(4)}$$

$$-\frac{1}{16}$$

7.) $\lim_{x \to 3} \frac{x^2-9}{x-3}$ Function, Factor, Conjugate, Try, L'Hospital's.

$$\frac{\cancel{(x-3)}(x+3)}{\cancel{(x-3)}}$$

$$x+3$$

$$3+3$$

6

Problems Limits

$[-\infty, -3) \cup [+4, +\infty]$

8.)

f(x) where *x=(-2)*

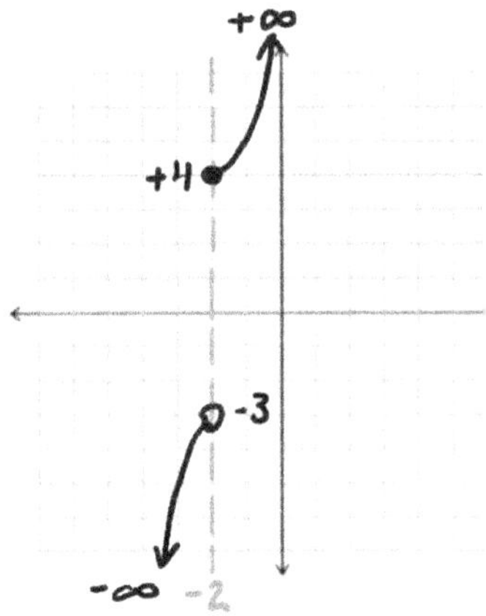

$$\lim_{x \to -2^-} = -3$$

$$\lim_{x \to -2^+} = +4$$

$$\lim_{x \to -2} = DNE$$

$$f(-2) \neq \lim_{x \to -2}$$

$\therefore$ *Not continuous*

Problems

Differentials 7.

1.) $y = f(x) = x^2$ D. Definition

2.) $y = f(x) = \sqrt{x}$ D. Definition

3.) $y = f(x) = 6x^3 + 5x^2 + 9$ Power

4.) $y = f(x) = x^3 secx$ Product

5.) $y = f(x) = \frac{secx}{x^4}$ Quotient

6.) $y = f(x) = \frac{1}{15}(3x - 2)^5$ Chain

7.) $y = f(x) = x^2y^3 - xy - 10$ Implicit

Problems Differentials 7

1.) $y = f(x) = x^2$ Definition

$$y' = \frac{(x+h)^2 - x^2}{\lim\limits_{h \to 0} h}$$

$$\frac{x^2 + 2xh + h^2 - x^2}{h}$$

$$\frac{2xh + h^2}{h}$$

$$\frac{\cancel{h}(2x + h)}{\cancel{h}}$$

$2x$

2.) $y = f(x) = \sqrt{x}$ Definition

$$y' = \frac{\sqrt{x+h} - \sqrt{x}}{\lim_{h\to 0} h}$$

$$\frac{\sqrt{x+h}-\sqrt{x}}{h} \cdot \frac{\sqrt{x+h}+\sqrt{x}}{\sqrt{x+h}+\sqrt{x}}$$ Conjugate

$$\frac{x+h-x}{h\left(\sqrt{x+h}\right)+\sqrt{x}}$$

$$\frac{h}{h\left(\sqrt{x+h}\right)+\sqrt{x}}$$

$$\frac{1}{\sqrt{x}+\sqrt{x}}$$

$$\frac{1}{2\sqrt{x}}$$

3.) Power

$y = f(x) = 6x^3 + 5x^2 + 9$

$y' = 18x^2 + 10$

4.) Product

$y = f(x) = x^3 secx$

$y' = 3x^2 secx + x^3 cosx$

$x^2(3secx + xcosx)$

5.) Quotient

$$y = f(x) = \frac{sinx}{x^4}$$

$$y' = \frac{x^4 cosx - 4x^3 sinx}{(x^4)^2}$$

$$= \frac{x^3 (xcosx)(-4sinx)}{x^8}$$

$$= \frac{xcosx - 4sinx}{x^5}$$

6.) <u>Chain</u>

$$y = f(x) = \frac{1}{15}(3x - 2)^5$$

$$y' = \frac{5}{15}(3x - 2)^4 \cdot 3$$

$$= \frac{15}{5}(3x - 2)^4$$

$$= (3x - 2)^4$$

7.) <u>Implicit</u>

$$y = f(x) = x^2y^3 - xy - 10$$

$$y' = \left[2x\frac{dx}{dx}y^3 + 3y\frac{dy}{dx}x^2\right] - \left[\frac{dx}{dx}y + x\frac{dy}{dx}\right]$$

$$= 2xy^3 + 3x^2y\frac{dy}{dx} - y - x\frac{dy}{dx}$$

$$= [3x^2y^2 - x]\frac{dy}{dx} = y - 2xy^3$$

$$\frac{dy}{dx} = \frac{y - 2xy^3}{3x^2y^2 - x} \cdot \frac{-1}{-1}$$

$$= \frac{2xy^3 - y}{x - 3x^2y^2}$$

Problems

Integrals 5.

1.) $f(x) = 4 - x^2$ $\int_{-2}^{+2}[f(x) - g(x)]dx$

$g(x) = x^2 + 1$ REVERSE.

$1_a^b = -2, +2$

FIND = AREA

2.) $\int (3x - 2)^4 dx$ U-SUB.

3.) $\int xe^x dx$ PARTS.

4.) $\int \frac{\sqrt{4-x^2}}{x^2} dx$ T-SUB.

5.) $\int \frac{X^2-1}{X^3+X} dx$ P-FRACTION

Problems Integration 5

1.) Reverse

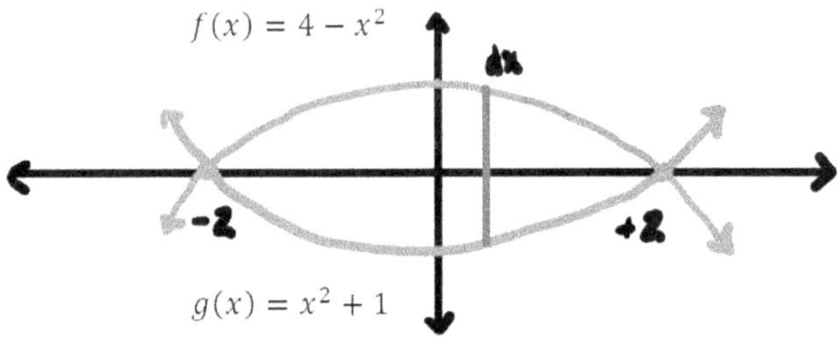

$$A = \int_a^b [f(x) - g(x)]\, dx$$

$\int_{-2}^{+2}[(4 - x^2) - (x^3 + 1)]\text{dx}$

$\left[\frac{-2(2)^3}{3} + 3(2)\right] - \left[\frac{-2(-2)^3}{3} + 3(-2)\right]$

$-\frac{16}{3} + 6 - \frac{16}{3} + 6$

$-\frac{32}{3} + \frac{36}{3}$

$\frac{4}{3}$

2.) U-SUB

$$\int (3x-2)^4 dx$$

$$u = (3x-2) \quad du = 3dx \qquad dx = \frac{du}{3}$$

$$\int u^4 \frac{du}{3}$$

$$\frac{1}{3}\frac{u^5}{5} + c$$

$$\frac{1}{15}u^5 + c$$

$$\frac{1}{15}(3x-2)^5 + c$$

3.) PARTS

$$uv - \int v\,du$$

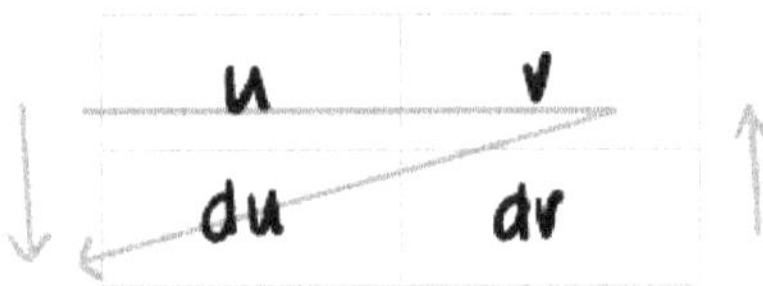

$$\int \underset{u}{\underline{x}}\,\underset{dv}{\underline{e^x dx}}$$

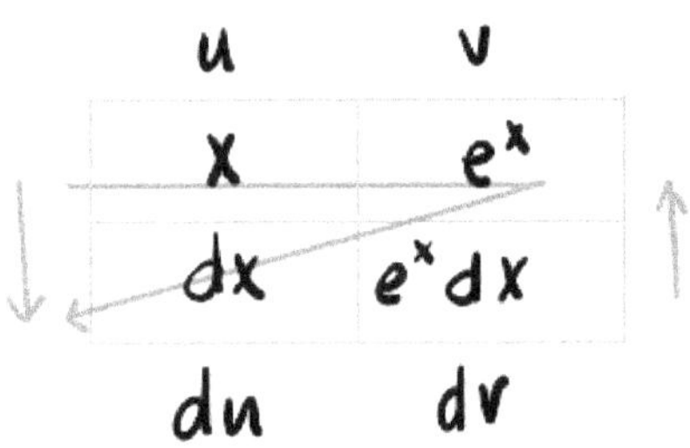

$$xe^x - \int e^x dx + c$$

$$xe^x - e^x + c$$

PARTS (2)

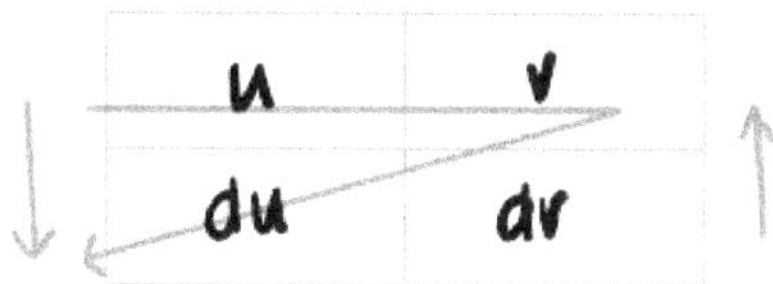

$uv - \int vdu$

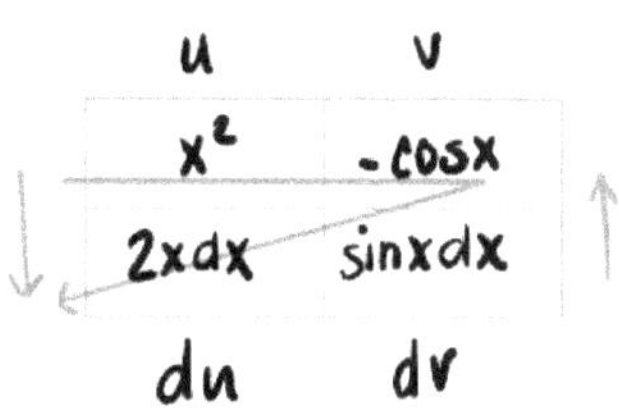

$$-x^2cosx + \int -2xcosxdx$$

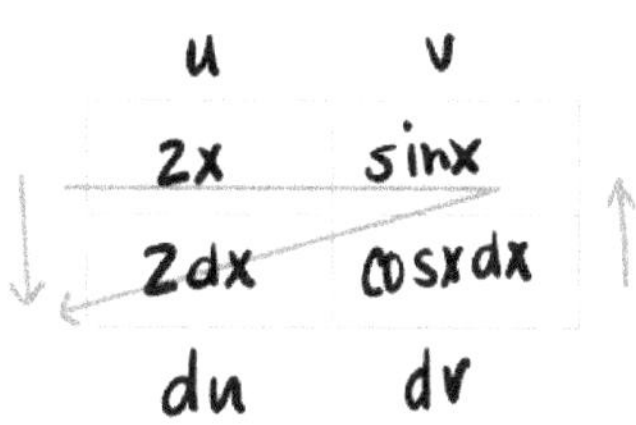

$$2xsinx - \int 2sinxdx$$

$-x^2cosx + 2xsinx + 2cosx + c$

$2xsinx + 2cosx - x^2cosx + c$

PARTS (SPECIAL)

$\int x^3 lnx dx$

$u \quad dv$

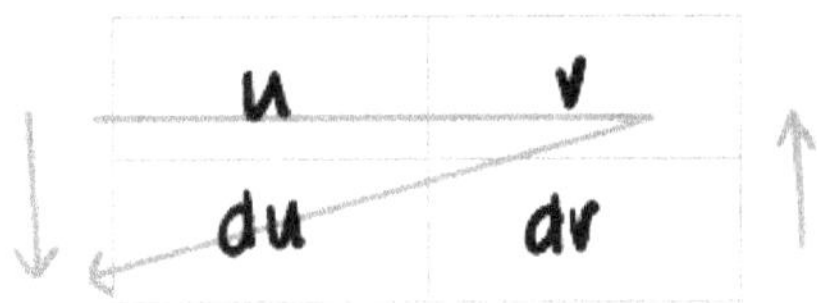

$uv -$

$\int vdu$

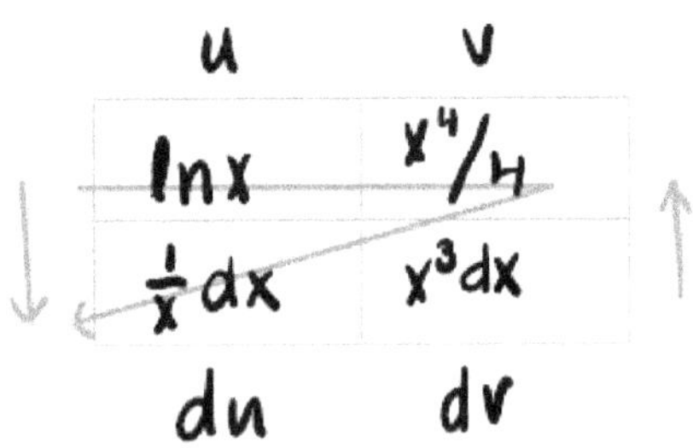

$$lnx\frac{x^4}{x} - \int \frac{x^3}{4}\frac{1}{x}dx +$$

$$\frac{1}{4}lnx(x^4) - \frac{1}{4}\left(\frac{x^4}{4}\right) + c$$

$$\frac{1}{4}x^4 \ln(x) - \frac{1}{16}x^4 + c$$

Problems – Integration

4.) <u>T-SUB</u>

<u>SISF</u>

SUBSTITUTE

IDENTITIES

SOH CAH TOA

FUNCTION

$$\int \frac{\sqrt{4-x^2}}{x^2} dx$$

<u>SUB</u>

$a^2 = 4 \quad a = 2 \quad x = 2sin\theta \quad dx = 2cos\theta d\theta$

<u>IDENTITIES</u>

$$\int \frac{\sqrt{4-4sin^2\theta}}{4sin^2\theta} 2cos\theta d\theta$$

$$\int \frac{\cancel{2}\sqrt{1-sin^2\theta}}{4sin\theta} \cancel{2}sin\theta d$$

$$\int \frac{\sqrt{cos^2\theta}}{sin^2\theta} cose\theta d\theta$$

$$\int \frac{cos^2\theta}{sin^2\theta} d\theta$$

$$\int \frac{1 - sin^2\theta}{sin^2\theta} d\theta$$

$$\int \left[\frac{1}{sin^2\theta} - 1\right] d\theta$$

$$\int [\csc^2 \theta - 1]\, d\theta$$

$-\cot\theta - \theta + c \leftarrow$

SOH CAH TOA

$x = 2sin\theta$

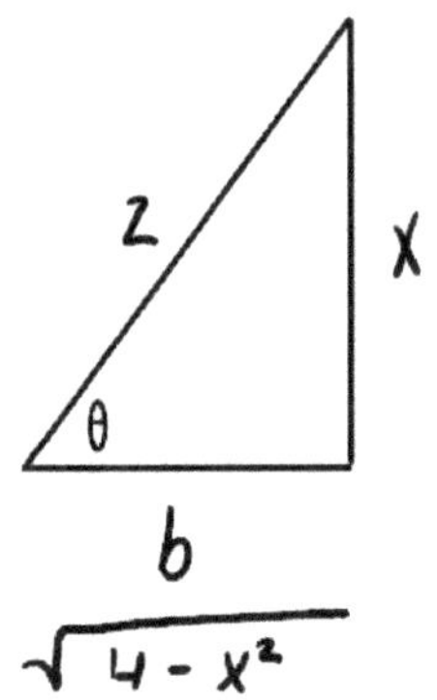

$sin\theta = \frac{x}{2}$

$b^2 = 4 - x^2$

$b = \sqrt{4 - x^2}$

FUNCTIONS

$-\cot\theta - \theta + c$

$-\frac{\sqrt{4 - x^2}}{x} - \sin^{-1}\frac{x}{2} + c$

Polynomial Division

$\frac{3x^3+2x-11}{x-3}$ Irrational $\left(\frac{x^3}{x}\right)$

$$
\begin{array}{r}
3x^2 + 9x + 29 + \frac{76}{x-3} \\
x - 3 \overline{\big) \; 3x^3 + 0x^2 + 2x - 11} \\
\underline{3x^3 - 9x^2} \qquad\qquad \\
9x^2 + 2x \qquad \\
\underline{9x^2 - 27x} \qquad \\
29x - 11 \\
\underline{29x - 87} \\
76 \textit{ remainder}
\end{array}
$$

$$\int \left[3x^2 + 9x + 29 + \frac{76}{x-3}\right] dx$$

$$\frac{3x^3}{3} + \frac{9x^2}{2} + \frac{29x}{1} + 76\ln|x-3|$$

$$x^3 + \frac{9x^2}{2} + 29x + 76\ln|x-3| + c$$

P-FRACTIONS

CRF(PD) L Q S G U/y'

$\int \frac{x-9}{x^2+3x-10}dx$

FT
-10
/ \
-2 5

$(x-2)(x+5)$

LINEAR DECOMPOSITION

$\left[\frac{x-9}{(x-2)(x+5)} = \frac{A}{x-2} + \frac{B}{x+5}\right]$ $(x-2)(x+5)$

SUB

$x - 9 = A(x+5) + B(x-2)$

$x = +2 : 2 - 9 = A(7) : -7 = 7A$ $A = -1$

$x = -5 : -5 - 9 = B(-5-2) : -14 =$

$B9(-7)$ $B = 2$

$$\int \frac{x-9}{x^2+3x-10}dx = \int \frac{-1}{x-2}dx + \int \frac{2}{x+5}dx$$

$2\ln|x+5| - \ln|x-2| + c$

P-FRACTIONS

CRF(PD) LQ SG U/y'

$$\int \frac{x^2-1}{x^3+x}dx = \int \frac{x^2-1}{x(x^2+1)}dx$$

$$\frac{x^2-1}{x(x^2+1)} = \frac{A}{x} + \frac{Bx+c}{x^2+1}$$

$$x^2-1 = A(x^2+1) + (Bx-c)(x)$$

SUB

$x = 0: \ -1 = A(1) + 0$ $A = -1$

GROUP

$$x^2-1 = Ax^2 + A + Bx^2 + cx$$

$x^2-1 = (A+B)x^2 + cx + A$

$x^2 = (A+B)x^2$

$1 = A+B$ $1 = -1+B$

$B = 2$ $0 = Cx$ $C = 0$

$\int \frac{x^2-1}{x^3+x}dx = \int \frac{-1}{x}dx + \int \frac{2x}{x^2+1}$

$\ln|x^2+1| - \ln|x| + c$

$\ln\left|\frac{x^2+1}{x}\right| + c$

U-SUB

$u = x^2+1$

$du = 2xdx$

$dx = \frac{du}{2x}$

$\int \frac{2x}{u}\frac{du}{2x}$

$\int \frac{1}{u}du$

$\text{Ln}|u|$

$\ln|x^2+1|$

PROBLEMS

AREA 2.

1.) $\int_a^b [f(x) - g(x)]dx$ TOB/BOTTOM

2.) $\int_A^B [f(y) - g(y)]dy$ RIGHT/LEFT

PROBLEMS

VOLUME 3.

1.) $\int_a^b a^2[f(x) - g(x)]dx$ SLICE

OR $[f(y) - g(y)]dy$

2.) $\int_a^b \pi r^2[f(x) - g(x)]dx$ DISK

3.) $\int_a^b \pi\Delta r^2[f(x) - g(x)]dx$ WASHER

WHERE: $\Delta r^2 = (R^2 - r^2)$

PROBLEMS AREA 2. (TBRL)

TOP/BOTTOM

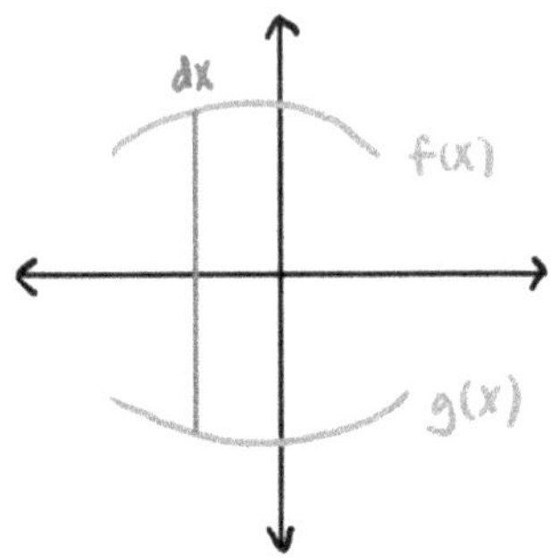

$A = \int_a^b [f(x) - g(x)]\, d$

RIGHT/LEFT

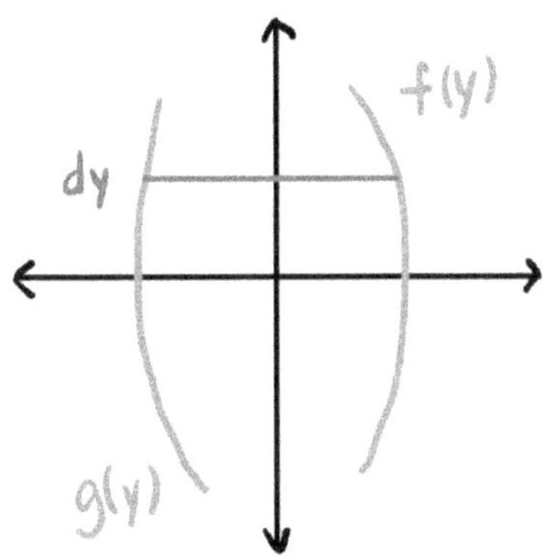

$A = \int_a^b \left[f(y_{-g(y)})\right] dy$

PROBLEMS VOLUMES 3. (SDW)

SLICE

$V = \int_a^b a^2 f(y)dy$ CIRCULAR SECTION

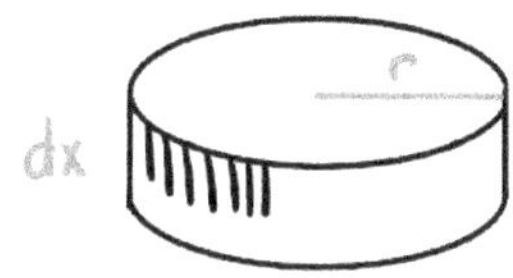

RECTANGULAR SECTION

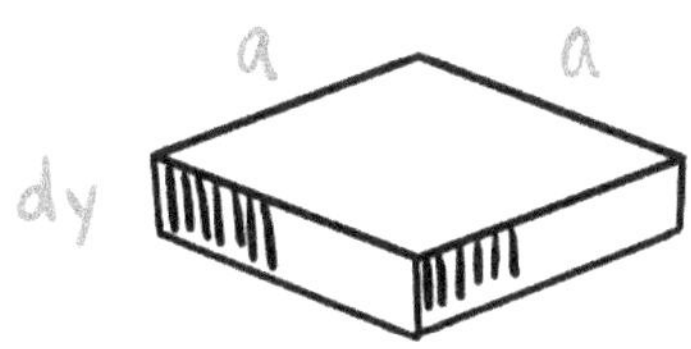

DISK

$V = \pi r^2 f(x)d$

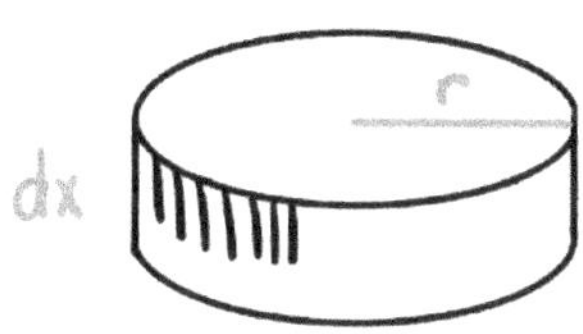

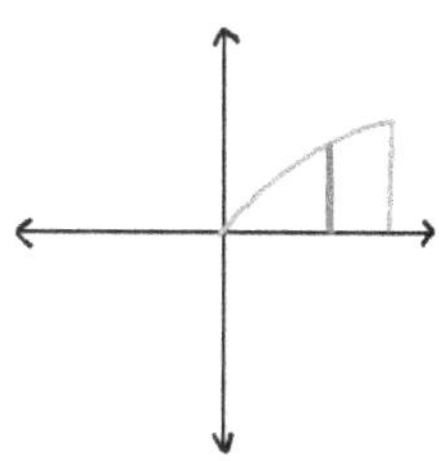

WASHER

$$V = \pi \Delta r^2 f(x) dx$$

WHERE $\Delta r^2 = (R^2 - r^2)$

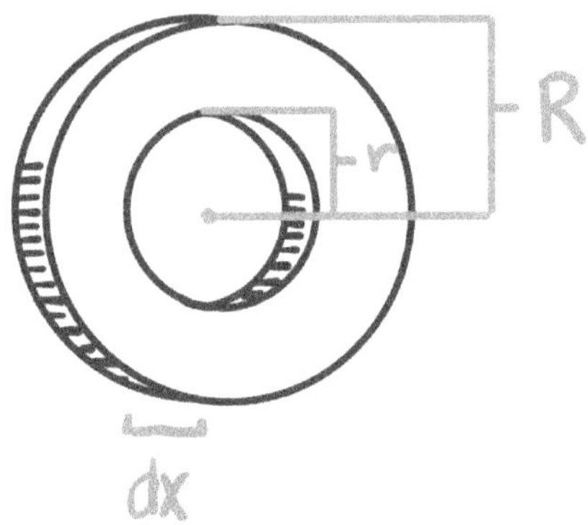

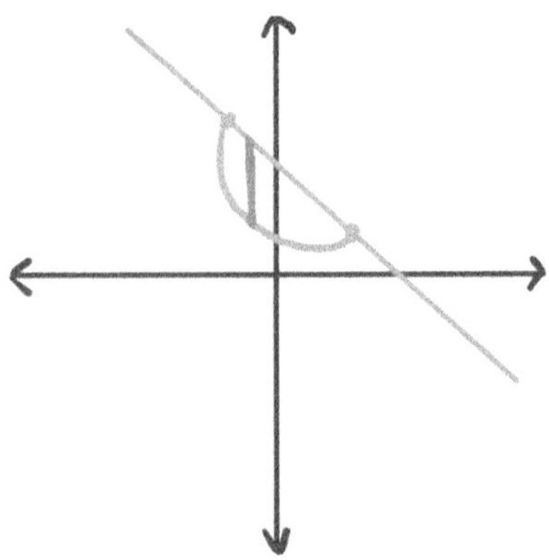

VOLUME CONE SLICE

SHAPE SECTION RATIO

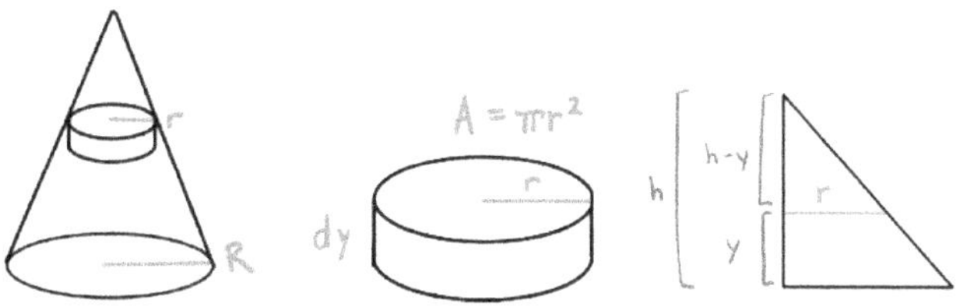

FORMULA

$$V = \pi r^2 h = \int_o^h \pi R^2 \left(1 - \frac{1}{h}y\right)^2 dy$$

U-SUB

$$u = \left(1 - \frac{1}{h}y\right) du = \left(-\frac{1}{h}y\right)\ dy = (-h)du$$

U-LIMITS

$u(h) = 0$ TOP $u(0) = 1$

BOTTOM

SOLUTION

$$\int_1^0 \pi R^2 u^2 (-h) du = \int_1^0 \pi R^2 (-h) u^2 du$$

$$= \pi R^2 (-h) \left[\frac{u^3}{3}\right]_1^0$$

$$= \pi R^2 (-h) \left[0 - \frac{1}{3}\right]$$

$$= \frac{1}{3} \pi R^2 h$$

SHELL $\int_a^b$

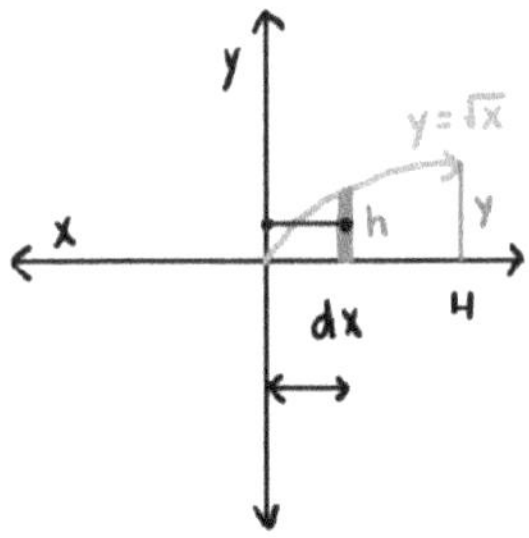

$\backslash y = f(x) = \sqrt{x}$

$R = xh = y$

LIMITS $0 \rightarrow 4$

$$V = 2\pi \int_a^b R(x)h(x)0dx$$

$$= 2\pi \int_0^4 x\sqrt{x}dx$$

$$= 2\pi \int_0^4 x^{\frac{3}{2}}\text{dx}$$

$$= 2\pi \left[\frac{x^{\frac{5}{2}}}{\frac{5}{2}}\right]_0^4 = \left[\frac{2}{5}4^{5-2} - 0\right]$$

$$= 2\pi \left[\frac{2}{5}4^3\right] = 2\pi \left[\frac{2}{5} \cdot 64\right]$$

$$= \frac{256\pi}{5}$$

SEQUENCE SERIES SUMMATION CONVERGE DIVERGE

<u>SEQUENCE</u>	<u>SERIES</u>	<u>SUMMATION</u>
$\frac{1}{2}, \frac{1}{4}, \frac{1}{8}, \frac{1}{16} \dots$	$\frac{1}{2} + \frac{1}{4} + \frac{1}{8} + \frac{1}{16} \dots$	$\sum_{x=a}^{b} f(x) + - \times \div$
COMMA	$+ - \times \div$	

<u>CONVERGE</u>	<u>DIVERGE</u>
$\frac{1}{2} + \frac{1}{4} + \frac{1}{8} + \frac{1}{16} \dots 0$	$2 + 4 + 8 + 16 \dots \infty$
A FINITE NUMBER.	INFINITY.

$\lim_{n \to \infty} a_n = \frac{1}{2^n} = \frac{1}{2^\infty} = \frac{1}{\infty} = 0$ CONVERGE

<u>SEQUENCE</u>

$\lim_{x \to \infty} S_n = \left(1 - \frac{1}{2^n}\right) = \left(1 - \frac{1}{2^\infty}\right) = 1 - \frac{1}{\infty} = 1 - 0 = 1$

CONVERGE

<u>SERIES</u>

$$\sum_{x=1}^{7} f(x) = x\Big|_1^7 = 1 + 2 + 3 + 4 + 5 + 6 + 7 = 21$$

<u>SUMMATION</u>

SEQUENCE SERIES SUMMATION

Arithmetic

<u>Sequence</u>

$a_n = a_1 + (n - 1)d$

<u>Series</u>

$s_n = \frac{n}{2}(a_1 + a_n)$

Geometric

<u>Sequence</u>

$a_n = a_1 r^{n-1}$

<u>Series</u>

$s_n = \frac{a_1(1-r^n)}{1-r}$ or $\frac{a_1 - a_n r}{1-r}$ Unknown Terms

Summation

$$s = \sum_{x=1}^{7} f(x)$$

Notes

$n = terms$

$d = a_2 - a_1$ or $n_2 - n_1$ Difference

$r = {}^{a_2}/_{a_1}$ Ratio

Converge Diverge

L'Hopitals

$$\lim_{x\to\infty}\frac{x^2}{2^x}=\lim_{x\to\infty}\frac{2x}{2^x\ln 2}=$$

$\frac{2}{\infty}=0$ Converge

Sequence

$\lim_{x\to 0} s_n = \lim_{n\to\infty}\left(1-\frac{1}{2^n}\right)=1-\frac{1}{\infty}=1-0=1$ Converge

Series

$\lim_{n\to\infty} a_n = \lim_{n\to\infty}\frac{1}{2^n}=\frac{1}{2^\infty}=\frac{1}{\infty}=0$ Converge

Note

<u>Converge</u>

$\frac{1}{2}+\frac{1}{4}+\frac{1}{8}+\frac{1}{16}\ldots$ Point

<u>Sequence</u>

$\frac{1}{2},\frac{1}{4},\frac{1}{8},\frac{1}{16}\ldots$ Comma

<u>Diverge</u>

$2+4+8+16\ldots\infty$

<u>Series</u>

$\frac{1}{2}+\frac{1}{4}+\frac{1}{8}+\frac{1}{16}\ldots+-\times\div$

Series <u>Limit</u>

L'Hopital's

$$\frac{{}^{d}/_{dx}\ Numerator}{{}^{d}/_{dx}\ Denominator}$$

Function y' y''

$$\lim_{x\to\infty} \frac{x^2}{2^x} = \frac{2x}{2^x \ln 2} = \frac{2}{2^x \ln 2 \ln 2}$$

$$= \frac{2}{\infty} = 0 \text{ Converge}$$

<u>Note</u>

${}^{d}/_{dx}\, 2^x = x \ln|2| = \ln|2| = \ln 2$ where $x = 1$

Converge = Finite

Diverge = Infinite

www.ingramcontent.com/pod-product-compliance
Lightning Source LLC
LaVergne TN
LVHW022204200926
848481LV00006B/1840